Studies in Systems, Decision and Control

Volume 300

Series Editor

Janusz Kacprzyk, Systems Research Institute, Polish Academy of Sciences, Warsaw, Poland

The series "Studies in Systems, Decision and Control" (SSDC) covers both new developments and advances, as well as the state of the art, in the various areas of broadly perceived systems, decision making and control–quickly, up to date and with a high quality. The intent is to cover the theory, applications, and perspectives on the state of the art and future developments relevant to systems, decision making, control, complex processes and related areas, as embedded in the fields of engineering, computer science, physics, economics, social and life sciences, as well as the paradigms and methodologies behind them. The series contains monographs, textbooks, lecture notes and edited volumes in systems, decision making and control spanning the areas of Cyber-Physical Systems, Autonomous Systems, Sensor Networks, Control Systems, Energy Systems, Automotive Systems, Biological Systems, Vehicular Networking and Connected Vehicles, Aerospace Systems, Automation, Manufacturing, Smart Grids, Nonlinear Systems, Power Systems, Robotics, Social Systems, Economic Systems and other. Of particular value to both the contributors and the readership are the short publication timeframe and the world-wide distribution and exposure which enable both a wide and rapid dissemination of research output.

** Indexing: The books of this series are submitted to ISI, SCOPUS, DBLP, Ulrichs, MathSciNet, Current Mathematical Publications, Mathematical Reviews, Zentralblatt Math: MetaPress and Springerlink.

More information about this series at http://www.springer.com/series/13304

Mikhail V. Belov · Dmitry A. Novikov

Methodology of Complex Activity

Foundations of Understanding and Modelling

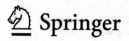

Mikhail V. Belov
IBS Company
Moscow, Russia

Dmitry A. Novikov
V.A. Trapeznikov Institute
of Control Sciences
Russian Academy of Sciences
Moscow, Russia

ISSN 2198-4182   ISSN 2198-4190   (electronic)
Studies in Systems, Decision and Control
ISBN 978-3-030-48612-9   ISBN 978-3-030-48610-5   (eBook)
https://doi.org/10.1007/978-3-030-48610-5

This Springer imprint is published by the registered company Springer Nature Switzerland AG
The registered company address is: Gewerbestrasse 11, 6330 Cham, Switzerland

Introduction

Rationale. There exist different types of human activity, from those of entrepreneurs developing high-tech businesses (Elon Musk, for example, 1a), engineers (George Stephenson, 1b), or government or public figures (Franklin Delano Roosevelt, 1c), to the activity of a gas station attendant (2a), an orange sorter (2b), or an assembly worker on the shop floor of an aircraft manufacturer (2c).

It is easy to see that some people's work activity is complex and diverse (1a-1c), whereas those of others are monotonous and routine (2a-2c). Yet the activity of the second group is also sometimes "complex": no one would claim that the worker who assembles the cockpit of a modern aircraft (2c) is not involved in "complex" activity. What they have in common is that they are all examples of human activity. But where do the differences lie? How do we formally determine the similarities and differences among different types of activity? One distinctive characteristic is the uncertainty of an activity—the uncertainty of external factors or the uncertainty of the goals, methods, and behavior of the actors involved. In 2a-2c, the activity is completely determinate, or seen to be completely determinate: the actors react to external uncertainties by referring the problem to a higher level actor and are responsible for carrying out activity within the prescribed procedures as directed from above, in contrast to 1a, 1b, or 1c, who react to uncertainty in the external environment and are not bound by predetermined goals and established procedures. Their reaction is to organize and to launch new activity, first and foremost to create new methods and procedures; they themselves are responsible for the final result of the activity, including the activity of subordinates. And why does the "indeterminism" of the activity of 1a-1c appear and what does it constitute? Again, it is easy to see that the activity of the former (1a-1c) is not always "equally complex." How do we delineate these differences in complexity in the activity of one and the same person? It is intuitively clear that not all elements of activity are "equally complex and uncertain." Moreover, implementing elements of activity with a high degree of complexity and uncertainty requires significantly higher costs (managerial and material resources and time) than do routine (less complex and uncertain) elements.

And what about management and organization activity? Is it possible to define specifically and formally what activity is included here? For example, a CEO may simply delegate tasks: what, then, is being managed? Is delegating a complex activity? And how about when the same CEO signs a contract? No one else has the right to sign, but the signing is almost a formality—based on previous discussions and others having already given their approval.

Current methodologies (the science of organizing activity) do not yet provide answers to these and similar questions. This book proposes a tool to do so.

Methodology of complex activity. This book introduces the idea of "*complex activity*"[1] as activity (where *activity* means meaningful human work or actions [97, p. 4]) with a non-trivial internal structure and with multiple and/or changing actors, methods, and roles of the subject matter of activity in its relevant context. In view of the distinctiveness of complex activity, it is taken into consideration here along with the implementing entity (as a rule, a Sociotechnical System (STS) [144]).

Consequently, the authors call the theory developed in this book—in the form of a series of assertions and an integrated system of models that constitutes a school of thought about the organization of complex activity—the *Methodology of Complex Activity* (MCA). In other words, the subject matter of this study is Complex Activity (CA), and the research topic is the general principles underlying its organization and management. MCA builds on a *general methodology* [97] as tailored to complex activity.

The proposed theory provides a systematic basis for solving such problems as follows:

- planning a new complex activity;
- considering alternative solutions in such a plan;
- developing procedural documentation;
- CA modeling, first and foremost computer modeling;
- creating CA management systems in the form of descriptions of management processes, specifications, knowledge and data used in corresponding information technology systems (both software and hardware), and, of course, trained employees;

and many others.

The models that make up the developed theory are diverse and quite numerous, which reflects the natural complexity of CA as a system, so modeling a large number of concrete elements of a CA is rather time-consuming and requires a great

---

[1]*Key terms in the text are shown in italics. If a well-known definition is used, the corresponding source is indicated.*

amount of effort. However, the system of models is constructed in such a way that it does not require an obligatory description of the "entire" complex activity each time. Such a system allows one to abstract and focus on the elements of interest and to model exactly these elements in detail, leaving the rest of the abstract as "black boxes" without losing the expressive properties of the models and without worsening the quality of the presentation.

Novelty. Due to "universality," activity of the kind that is the thrust of this book touches, to some extent, upon many areas of knowledge. Accordingly, the methodology of complex activity is associated with many scientific disciplines and many intellectual traditions, so it makes sense to immediately clarify the novel elements of the proposed approaches.

In this book, first, unified means are proposed to formally describe and analyze any complex activity, along with the players involved "in the entire range from 1a-1c to 2a-2c."

Second, the role of uncertainty is analyzed. It is shown that the complexity of an activity comprises manifestations of uncertainty (uncertainty in goals, uncertainty in results, uncertainty in external conditions, etc.) and how to deal with it. Uncertainty comes to be through the onset of a priori unpredictable events, and the reaction to this may be a new activity that was not present before their onset.

It is shown that complex activity, in spite of its intricacy, is basically "mechanistic"; it is "predetermined" a priori as determinate. In the overwhelming majority of examples of complex activity, uncertainty exerts an influence on them, but the reaction to uncertainty is formulated outside the activity: faced with a problem, the assembly worker at the aircraft manufacturing enterprise has no right to retool the work methods and procedures or to make changes. She is obliged to strictly comply with procedural norms and as a matter of course to refer problems to her superiors and then to the engineers. Thus, apparently "99%" of activity is "mechanistic," and the remaining "1%" is actually "complex"; it is shown below what this complexity is and how it is manifested.

Third, such activity as *management* and *organization* (as processes) is formalized and investigated. The components of organization—*analysis, synthesis,* and *concretization*—are ascertained and studied, as are the components of *management*: *organization, regulation,* and *evaluation*. It is shown that the focus of organization and management in relation to complex activity is the amalgamation (system of systems) of complex activity and the entity that implements it (the sociotechnical system).

Fourth, the role of *technology* in an activity is identified: activity connected with the development of technology is indeed "complicated," while all other activities, including organization and management, are routine! Management and organization become "complex" when, due to uncertainty, in the course of their implementation it is necessary to develop methods and tools (technology) for a new activity, because the existing ones are insufficient to adequately respond to uncertainty.

Back in History. The table below describes the organizational specifics of activity during different periods of human development, from the Stone Age to the present day. The main conclusions that can be drawn from such a periodization are as follows:

1. Starting from the appearance of man, human activity is characterized by multiple and changing goals and other attributes of the definition of CA. In other words, human activity has always been complex and MCA provides a uniform description for CA in all periods. Unlike other living creatures acting jointly (a swarm of bees, a colony of ants, a pack of wolves, etc.), man organizes all types of CA (labor, learning, play, creativity, communication), i.e., performs structuring of the subject matter, in accordance with the goals of CA not instinctively but consciously, as a form of management. The second distinctive feature of human CA is the use of artificial means of activity.

2. Besides the monotonously growing complexity of CA (the depth and width of its hierarchical logical structure), technologies have been the only evolving factor of CA. But even for technologies, it is difficult to identify certain "historically specific" forms and methods of activity. The means of activity have been evolving! That is, in the course of human development, the means of CA and methods for performing "industry-specific" elementary operations (industry-specific technologies) have been created and further developed. At the system-wide level of generalization, the technology of CA has been remaining invariant: the target (logical) structure and the cause-effect structure, as well as the process model of CA as a universal algorithm for managing and/or implementing the life cycle of CA.

Characteristic	10000 B.C.	5000 B.C.	0	500 A.D.	1000 A.D.	1500 A.D.	2000 A.D.
Social structures (K. Marx)	Primitive-communal system		Slave system		Feudalism	Capitalism	... Communism
Types of organizational culture (V.A. Nikitin, A.M. Novikov)	Traditional			Corporate-handicraft		Professional	Project-technological ... Knowledge
Mass types of practical activity	Hunting, fishing, collecting	Cattle-breeding, agriculture		Craftsmanship		Industrial production	Information production
Sources of energy	Muscle force of people and animals			Natural sources (water, wind)		Hydrocarbons (steam, electricity)	Nuclear energy, RES
Dominating types of production		Piece			Batch	Mass	
Methods of normalization and translation of activity	Myths and rituals		Sample and recipe for its recreation		Theoretical knowledge in the form of text	Projects and programs	Information models
Organizational forms of collective activity	Community		Church		Workshop	Enterprise	Corporation ... Extended or virtual enterprise
Dominating links between actors performing joint activity	Kinship	Language	Faith	Property		Capital	Organization ... Technology

... 10000 B.C. ... 5000 B.C. ... 0   500 A.D.   1000 A.D.   1500 A.D.   2000 A.D.   **Time** →

3. The maximum complexity of the projects implemented in different historical periods (which can be described, e.g., by the spatial dimensions of objects created and the number of their "elements," the duration of projects and the number of their participants) has been demonstrating a moderate growth over time.
4. The gradually accelerating development of technologies has led to the mass creation of more and more complex artificial systems, which allows achieving results with fewer resources (time, energy, etc.). In fact, "the set of achievable results" over the entire history of mankind has not changed much (a few exceptions are hydronautics, astronautics, etc.).

Thus, the typical architecture of CA put forward by MCA is universal for any types of activity, throughout the entire past history of mankind. Hence, it can be hypothesized that such universality will hold in the future (at least, until the set of subject matters of activity that is accessible to mankind changes).

Main Results. The results obtained are formulated as a unified theory that provides a description and examination of complex activity, organization, and its management (as processes). The theory consists of a series of assertions and an integrated set of common models. Based on a fundamental understanding of the methodology [97, 98, 99], practical observations of how complex activity is implemented, and logical approaches, a number of conclusions reflecting the logic of the development of the theory and the components that make it up are subsequently formulated. The model system plays the role of a *framework* constituting a tool for solving practical and theoretical management challenges. This system of models allows modeling the "system of interest" and describing in an aggregated way the external environment for it with the necessary detail.

This allows not only structuring complex activity, but also reasonably breaking down the elements of an activity by the degree of complexity and uncertainty, singling out the most critical ones, respectively, in practice—allocating resources and managerial and organizational efforts. By the same token, such capabilities of the proposed system of models provide an opportunity for optimization, i.e., solving management problems in relation to CAs—a synthesis of effective processes of organization and management of complex activity and its players—via complex systems.

Structure of the Presentation. Chapter 1 is devoted to laying out the task of developing a methodology of complex activity. The relevance of the issue is analyzed, a general logical research plan is outlined, and the research topic is defined: complex activity as the primary one and the sociotechnical system as a secondary research topic.

In Chap. 2, definitions of elementary and complex activity are introduced and the defining features of CA are analyzed. The requirements for a methodology of complex activity are laid out, and a comparative analysis of the related branches of knowledge on the question of satisfying the requirements for MCA is carried out, that is, the necessity of developing MCA as a new theory is substantiated.

In Chap. 3, an analysis of the structural features of CAs is performed, and a unified formalism is proposed for describing the *Structural Element of Activity* (SEA) as an integrated subject, which describes the elements of CA and defines the rules for operating SEAs. The logical and cause-and-effect structures of CA are introduced as a set of links between SEAs.

Chapter 4 is devoted to questions on the origin of various kinds of activity; it introduces the classification of CAs and SEAs, and a generic model is proposed for the realization of activity (behavior of SEAs and CAs as a whole).

Process models of complex activity, its execution, and the fulfillment of the life cycles of its various elements are given in Chap. 5.

Chapter 6 introduces the metrics of complex activity: indicators and criteria for its effectiveness and performance. The system-wide factors on which the effectiveness and performance of CAs depend are discussed.

Chapter 7 is devoted to such types of CAs as organization and management, including CA optimization questions.

The appendices comprise the following: main abbreviations (Appendix 1), main definitions (Appendix 2), main assertions (Appendix 3), and results of analyses of the interrelationship between MCA and its requirements (Appendix 4). The assertions in the text are featured in light gray boxes, with a line separating the title in bold type from the description.

Several typical examples of CA are used to illustrate the implementation of common approaches throughout the text, referring to spheres of human activity that differ significantly from each other: the functioning of work groups, organizational units, projects, and organizations in general:

- a retail bank,
- an aircraft manufacturer,
- a fire department, and
- a nuclear power plant.

We would like to suggest several options for becoming familiar with this book. The first—and the most superficial—is to read the Introduction, Sect. 1.1, the conclusions in Chaps. 1–7, Sects. 7.1 and 7.2, and the Conclusion. A more detailed level of understanding will require additional familiarization with the material of Chaps. 2–5. Finally, for a complete picture, it is best to read the entire book in order.

The authors are grateful to V. N. Burkov, A. A. Voronin, A. O. Kalashnikov, G. N. Kalyanov, V. V. Kondratiev, N. A. Korgin, R. M. Nizhegorodtsev, V. V. Novochadov, A. N. Raikov, S. A. Saltykov, A. G. Teslinov, and G. L. Tsipes for their discussions, criticism, and valuable comments.[2]

─────────────────────
[2] *The authors are also grateful to the Russian Science Foundation (grant no. 16-19-10609) for partially supporting this research.*

Contents

Chapter 1
Formulation of Problems and General Approaches

This chapter serves as the introduction. The first section is devoted to presenting the details of the research topic; in the second section, the relevance of the question of creating a methodology of complex activity is analyzed; and in the third section, a general, logical system for constructing such a methodology is laid out.

1.1 Research Topics

Complex Activity and Sociotechnical Systems. Activity as meaningful human work and actions is just as universal a part of human existence as the satisfaction of basic needs. However, unlike the latter, complex activity is quite intricate, as are the actors involved in it. In the modern world, the bulk of the gross output (as a result of human activity) is created in enterprises, in companies, in organizations, within projects, in planning bureaus, in state, regional, and municipal agencies and entities, and in transnational corporations and their subsidiaries, as well as in various associations and groupings of all of the above entities and operations, along with all kinds of information and technical assets, systems, and equipment associated with them. All these entities and operations tie together several fundamental notions. First, they constitute *complex systems*[1]; second, they include *people* as elements, and third, a significant number of their constituent parts are *artificial*, that is, they are human made. By

---

[1] *System (from the Greek* whole, composed of elements, integrated*): an assemblage of elements that are related to and connected with each other and form a distinct totality, a unity. "A system (artificial) set of interacting elements, organized to achieve one or more declared goals. ... The system (in the applied sense) is often considered as a product (activity) or as a service that the system provides" [67, p. 9].*

Complex System: a system possessing the property of emergence; an open system with continuously interacting and competing elements. Openness is understood as free and unlimited (by artificial factors) participation and interaction of elements with each other and the environment [154].

© The Editor(s) (if applicable) and The Author(s), under exclusive license
to Springer Nature Switzerland AG 2020
M. V. Belov and D. A. Novikov, *Methodology of Complex Activity*, Studies in Systems, Decision and Control 300, https://doi.org/10.1007/978-3-030-48610-5_1

combining similar entities according to a specific rationale, they can be designated as *socio-technical systems*[2] (STSs), defined as complex[3] systems involving people, and, perhaps, technical and natural elements.

The concept of STSs encompasses almost all systems used and created as a result of human activity and which are made up of people. On the one hand, STSs are the results and subject matter of human activity; on the other hand, human activity is carried out within the framework of STSs: STSs function as complex actors in activity, that is, STSs are the actors, subject matter, and/or means of complex activity. Let us clarify this statement.

First, being organized and purposeful, practically all human activity is carried out within the framework of one STS or another, each of which has its own behavior concerning existence and development.

Second, STSs do not represent value in and of themselves; they do not bring benefits simply through their existence. Moreover, as material objects, they always require expenditures to maintain them. Value and utility arise from their implementing objectives, from their carrying out activity, from their functions. In this regard, socio-technical systems play the role of the *actors involved in activity*, for which they are actually created. Likewise, neither management in and of itself nor management systems are intrinsically valuable: from the point of view of the end result, they are not needed, nor is even the management impact on the object, but the value represents the state of the object being managed which was achieved as a result of a management exercise. That is, both STSs and their management constitute unavoidable "costs" when achieving the ultimate goals.

Third, STSs are the *results* and/or *subject matter* of another activity, and, being complex, they require a suitably complex activity from which to come forth. To correctly and completely describe such an activity, it is necessary to expand the concept of activity as currently accepted in philosophy, psychology, and methodology and to introduce a definition of complex activity. The understanding of *"complexity"* is discussed along with that of uncertainty and "emergence" in Sect. 2.7 below; here, we merely note that the complexity of the activity (especially a non-routine one) is usually no less than that of its subject matter.[4]

Emergence: that property of a system consisting of the fact that the properties of the whole are not reducible to the totality of the properties of the parts from which it is made up and are not derived from them.

[2]*The definition of STS corresponds to the fairly common term Enterprise System [116]. This definition extends the definitions of technical, organizational, ergatic (involving the human element), and socio-technical [114] systems. As a rule, an STS also includes the technological measures and technical components.*

Technology is a system of conditions, criteria, forms, methods, and means of successively achieving the defined goal (defined by the authors based on [97, p. 44]).

Technical components are an ensemble of artificial means of activity.

[3]*More precisely, a very complex system as defined by Beer [7].*

[4]*The validity of this assertion is substantiated—regardless of the definition of complexity—as follows. The complexity of the system does not decrease monotonically with regard to the number of elements and interconnections within the system. The creation of the subject matter of activity as a system requires the creation of all its elements and systematizing them—establishing links*

1.1 Research Topics
3

Fig. 1.1 Activity and its actors and subject matter

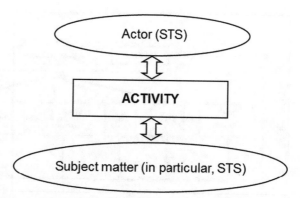

Thus, complex activity and the sociotechnical system together form a system of systems, a complex *dialectical pair* with each opposing the other (see Fig. 1.1): "the actor (STS) vs. the activity" and "the activity vs. the subject matter (in particular, STS)."

This book is devoted to the study of this pair: the complex activity and the sociotechnical system.

Complex activity, organization, and management. The vast majority of artificial systems has a goal and therefore requires organization and management. Purposefulness means result-oriented, aiming to obtain the desired result, achieving certain goals. The source or root cause of any result is activity, with its components and elements, and, above all, people as the actors involved in activity, the most important component of the activity. Organization and management[5] represent methods of influencing a managed/organized object to achieve defined goals. Insomuch as the "fountainhead" of any result is the corresponding activity, influence must be exerted on the activity and its components.

To lay out the subject matter of this study, it is necessary to analyze the categories of organization, management, and complex activity and consider how they relate to each other and to the category of sociotechnical system.

According to the Webster Dictionary, *organization* is defined as:

---

between them. Then each created element and each established connection must correspond to its own element of activity, which must also be ordered, that is, interrelated. Therefore, the number of elements and connections in an activity are no fewer than the number of elements and connections in their subject matter.

[5] *In this book, management is defined as follows: "a complex activity ensuring the impact made by a manager (the participant in this CA) on a system being managed (the management target) designed to ensure that his/her behavior leads to the achievement of the entity's objectives." The behavior of an object and/or a system is generally understood to be a successive (in time), at least partially observable, responsive, measurable objective fixation of its state changes. For an individual, behavior is a sequence of his/her actions, an interaction with the environment, mediated by his/her external (motor) or internal (mental) actions.*

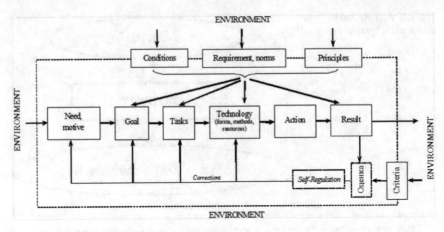

Fig. 1.2 Procedural components of activity [97]

(1) internal *orderliness, consistency* of interaction of more or less differentiated and autonomous parts of a whole, determined by its structure (organization as *property*);

(2) a combination of processes or actions leading to the formation and improvement of *interrelations* among parts of the whole (organization as a *process*, as an *activity*);

(3) the unification of people who jointly implement some program or goal and act on the basis of certain procedures and rules (organization as an *organizational system*).

To organize activity means to arrange it into a coherent system with clearly defined characteristics.

The general *methodology* considers activity in the form of a set of *procedural components* (see Fig. 1.2 [97, p. 31], generalizing and structuring Aleksei Leontev's model [76].

The *procedural* components of the activity [97] are as follows: the need, goal, tasks, and technology (including the forms, resources, and methods involved in the activity); and the impact and results of the activity (the "process" links among them are shown in bold arrows in Fig. 1.2). The following *characteristics of the activity* are external to this set: special features, principles, criteria, conditions, and norms. This understanding and the design of the methodology made it possible to create and lay out methodologies for scientific research [98], utilitarian and artistic activity, educational and interactive activity [97], and management [99] from a single standpoint using a uniform logic. The *actors involved in the activity* (who carry out the activity) and the *subject matter of the activity* (which changes throughout the course of the activity) are the key categories without which a description of the activity is undoubtedly incomplete.

Let us consider how the activity and its actor correspond. Without question, the following dialectical dualism is valid: without an actor, an activity can not exist (or be carried out) and, at the same time, an individual or STS can not be an actor outside an activity ("without an activity"). It is undoubtedly expedient to be guided by the following principle: there is at least a short initial period during which the CA exists in an informational form of a *desire for a solution*, an idea "in the head" of one (!) individual (for now an individual, not artificial intelligence and, that being so, the individual "creator" of a primary idea can always be unambiguously identified); we will call them the *"initiator"*. In this period, after the initiator has formulated the internal desire for a solution but has not yet involved anyone in the discussion or realization of the CA, the actor involved in the activity, as an example of a sociotechnical system, consists of that person alone.

It is important to note that the "initiator's" idea (need) did not appear when he/she was already carrying out a new activity, but namely when they were implementing some other activity[6] (or were "inactive": the lack of activity can be regarded as its particular case).

With further implementation, the activity becomes more complicated; it has a substantial internal structure, and other actors with their own know-how vis-à-vis the activity are drawn in. Accordingly, the generalized actor of this activity becomes more complicated: they actually become a sociotechnical system.

Therefore, it can be said that activity in the form of intention-need arises no later than its actor (the activity's actor!) in the form of a single individual, an elementary STS. In other words, the actor as an individual (*elementary STS*) and activity, as a minimum, "emerged equally early on." But activity in a more complex form is primary with respect to a non-elementary STS, since a complex STS is formed in response to the need to implement the activity.

> **Primacy of complex activity vis-à-vis the actor**
>
> Activity in the form of intention/need arises no later than its actor does as a single individual, an elementary STS. At the same time, a complex activity is primary in relation to its actor, a non-elementary STS.

This assertion can be illustrated by the following examples.

(a) A client coming into a branch of a retail bank to pay utility bills or open a savings account engenders bank activity: the provision of a specific service. First, the bank employee determines the client's needs. Since the range of services is limited, business process rules and documents that determine the makeup of the actors involved in a specific activity and the know-how to implement it have been previously developed for each service. A chain of business processes (actions) corresponding to

---

[6]*Furthermore, there is some continuity in time and the dynamics of how need is established in relation to the fact that, in the process of creating STSs to fulfill the need, they, by definition, interact and influence each other. Therefore, the need established undergoes an adjustment at the initial stages of the life cycle of the STS. Once the adjustment to the need comes into conflict with the existing STS, it acquires the features of a "new need," creates a "new STS," etc.*

these documents is implemented, with the immediate composition of the participating employees (the makeup of the actors) being determined in the course of carrying out this activity.

(b) The need to develop a unit and possibly the technology to manufacture or assemble parts during the design or modernization of an aircraft model engenders the appropriate activity. At some point, the need will arise for one of the designers to develop a unit that is part of the assemblage/working group developed by that designer. In accordance with the regulations, that person draws up a design request and passes it on to a coworker authorized to organize and plan new work. This coworker initiates the release of organizational and planning documents to carry out the new work. In this regard, the know-how and the makeup of the personnel may be off the shelf or specially created to perform a unique task.

(c), (d) When a fire department receives a call or equipment failure is discovered at a nuclear power plant, typical types of activity spring up and personnel (crews, teams, working groups) of a typical composition are put together.

Another argument in favor of the primacy of CA with respect to the actor and STS is that it is exactly the activity that provides the desired benefit, while the sociotechnical system cumulates the costs to maintain it.

A significant practical conclusion follows from the theoretical observation about the primacy of the CA with respect to the non-elementary STS: in analyzing, creating, and managing sociotechnical systems, the focus must not be on the STS itself, but on the complex activity created or acquired to implement the STS. That is, the kind of "secondary" firms, organizations, project teams, enterprises, and—even more so—productive and non-productive assets in relation to the activity (goals, results) that they are called upon to carry out must be kept in mind. In fact, a complex activity establishes requirements for the STS, which is its actor.

For many modern enterprises, activity is a *system-forming factor*. It is no secret that many organizations have functioned—both previously and currently—for the sake of self-existence, carrying out, for example, the search for business in order to "feed" employees, utilize equipment, etc. A more striking example is bureaucracy, which itself often "dreams up" activity for itself and others (subordinates, the population, etc.) in order to justify its existence and ensure its own growth.

Let us consider how the categories of activity and management relate.

Subject matter of organization and management in relation to complex activity

In managing and organizing a CA and/or STS, the subject matter is complex and includes a complex interacting and interrelated pair: "CA ⇔ STS".

Complex activity is the primary subject matter of organization and management, and the sociotechnical system itself is an intermediary, playing the roles of the actor involved in the CA and/or its subject matter.

A detailed consideration of management problems requires a more detailed analysis of the management category and its correlation with related concepts—first and foremost, with the organization category.

The definition of **management** as a *complex activity that ensures the impact*[7] *of a manager (the actor involved in this CA) on the system being managed (management target) designed to ensure his/her behavior leading to the attainment of the objectives of the manager* highlights several aspects.

First, it is stated that there is an actor making an impact. Sociotechnical systems are the managers in this study; therefore, the impact on the management target is the result of the activity of these STSs. Thus, *management is a special case of activity* (just as in the framework of the management methodology, management is a special case of practical activity [99]). In view of the complexity of the actor and the target, management, as a rule, constitutes a complex activity.

Second, by virtue of the specifics of sociotechnical systems, the actor involved in managerial activity is a complex dialectical pair {STS and CA}. It is important that the management impact be directed toward a system that is organized and focused, meaning the *management subject matter (as an activity) is generally complex and includes a pair {a sociotechnical system and complex activity}.*

Third, both the definition itself and the first of these noted aspects (management is activity, and activity is always purposeful) emphasize the *purposefulness of management.*

Fourth, the definition refers to the direction of the impact on the management target, which in turn ensures that the result required by the manager is obtained. Put another way, *management activity affects the final result indirectly*[8] *and implicitly through the management target.*

Fifth, the emphasis is on "ensuring the behavior" of the management target, i.e., *managerial activity is aimed at changing*[9] *the sequence of states of the management target* (see the definition of behavior of an object and/or a system in the Introduction.

From the point of view of an "external observer," the state of any real-world object can change over time for two reasons. The first is derived from some internal properties of the object and under the influence of the (possibly un-targeted) external environment; we will call this change "*inherent behavior.*" Second, the change in the state of the object can occur under the influence of this observer, who has their own goals and tries to influence the object. However, the observer's influence on the object is activity of the first, meaning that the reasons for any changes to the state of the object from the viewpoint of the observer are the inherent behavior of the

---

[7]*The term "impact" is used as a designation of the result of the corresponding (managerial) activity and can be construed as* management in the narrow sense *(see also Chap. 7). For example, if we are talking about management in a technical system—about automated management—the impact made by a technical control device (by a "pseudo-actor") is the implementation of the management algorithm established by the developer.*

[8]*For example—under* self-management—*the actor and the management target can "coincide."*

[9]*Any activity is performed in order to produce the desired changes (preserving the state is an example of change). Changes may not be implemented, but their goals must be defined. Otherwise, the activity becomes purposeless, that is, it ceases to be an activity.*

object (naturally, including changes in the behavior of the object under the influence of results of activity by other actors apart from this one) and/or the activity of the observer him or herself.

The above aspects of management categories, considered to be activity, allow it to be correlated with categories of organization, which is also considered to be an activity.

In this book, the following definition is employed: *an organization is a complex activity*[10] *aimed at creating internal order and co-ordination among the interactions of more or less differentiated and autonomous elements of the actor involved in this activity (including through the formation and maintenance of interrelationships with given characteristics among these elements).*

Thus, we come to the conclusion that *both management and organization are an activity.* As a result, it makes sense to highlight the similarities and differences between organization and management as activity in relation to the case considered in this book, where the actor involved in the activity is an STS.

The first three aspects highlighted above are inherent in both management and organization, and represent an activity (usually complex), the subject matter of which is a complex dialectic pair (STS and CA), and which is purposeful.

The fourth aspect, the *indirectness and obliqueness* of influence on the final result (the result of the activity of the management target) through the *subject matter of this activity* (the *management target*) constitutes the distinguishing feature of "management" and allows it to be identified from among all the elements of the "complex activity" array.

Analogously, *the goal in the form of a change in internal order*—the co-ordination of interacting parts—unambiguously singles out organization from among other forms of activity; therefore, it is the distinguishing feature of organization.

Essence of organization and management

Both management and organization (as a process) are activity, usually complex one, the subject matter of which is the CA and the corresponding STS.

At the same time, the indirectness and obliqueness of the influence on the final result through the subject matter of management activity is a distinctive feature of "management," and the goal in the form of a change in internal order and coordination of interacting parts is a distinctive feature of "organization."

A more detailed understanding of organization and management from the point of view of the MCA is discussed in Chap. 7, where it is shown that *organization* includes such types of activity as *analysis, synthesis, and concretization,* while *management* includes *organization, regulation, and evaluation.*

---

[10]*As in the case of management, an organization in the narrow sense can be construed to be a result or specific actions implemented within the framework of this organizational activity (see Chap. 7).*

An investigation of the organization and management of CAs and STSs using the MCA allows the assertion that, in comparison with general methodologies [97], which is for the most part *descriptive* in nature and investigates how an activity is organized, the next step below is *normative*, i.e., the issues of organization and management are studied and it is determined how to manage and organize activity.

The definitions of organization and management as "activity overseeing activity" possess a *recursiveness* (Fig. 1.3) and, as a result, there is the potential for an infinite expansion of the subject matter.

Organization and management, looked at as activity, are inherent in the actors and, in either case, the activity along with the corresponding actors can also be considered as a matter of "superior" activity in the organization and/or management (strictly speaking, they must be considered, since any activity has to be organized). In addition, this superior activity has an actor and so on. No activity subject matter can be so in the absence of activity, while activity can not exist without its subject matter. An equivalent assertion also holds for the actor involved in the activity. Therefore, "activity overseeing activity"—being recursive and nested—is potentially infinite. Consequently, attempting to include the "entire superior" recursive chain of subject matter and actors in the subject matter is not constructive.

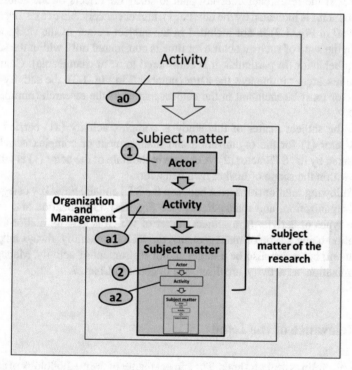

Fig. 1.3 Illustration of the definition of research subject matter

Fig. 1.4 Key categories of
MCA

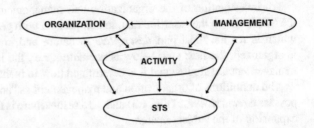

However, the subject matter of our study—and of every specific study of CA—is naturally limited (see the principle of abstraction and generalization in Sect. 1.3), due to the following considerations. It was noted above that either the behavior of the object itself or the activity of the very observer constitutes the reason behind any changes in the state of an object from an observer's point of view. Therefore, the condition for inclusion of the actor—who is superior in the recursion—into the research topic framework can be determined by the answer to the question, "Is the given actor considered to be changing, following only from their own behavior, or does the researcher also intend to study the influence on the superior actor's behavior?" If the researcher does not plan to study the effects on the actor (in the figure, this actor is indicated by the number 1), higher conceivable or existing activity (number a0 in Fig. 1.3) is not included in the subject matter of the study, and the change in the state of such an actor over time is considered only within the scope of their own behavior (in particular, it is considered to be unchangeable). Conversely, when the researcher influences the actor (number 2 in Fig. 1.3), the activity-impact on the actor must be included in the subject matter of the research (number a1 in Fig. 1.3).

Thus, the subject matter of this study is complex activity (a1) carried out by the STS/actor (1) for the organization and management of complex activity (a2) implemented by the STS/actor (2). A change in the state of the actor (1) is taken into account within the scope of his/her own behavior.

The following relationships exist between the STS and the three key categories—activity, organization, and management (see Fig. 1.4). Organization and management are types of activity (the subject matter of which is activity and/or the STS, organization being a type of management); the STS is generally always a "part" of management, but it can also be a target and/or instrument of activity. Management and organization, as activity, are discussed in detail in Chap. 7.

1.2 Relevance of the Issue

Related disciplines and methods. The subject matter of the methodology of complex activity (general patterns of organization and management of CA and STSs) is closely related to the question of managing complex organizational and technical systems, which has long been a research topic (see the review and references in [99]). First and foremost have been philosophers, since this issue is interdisciplinary, and this was realized from the very first attempts to resolve it, simply because it "did not fall, did not fit" (in the era of post-differentiation of the sciences) into the scope of interests of any of the "mono-disciplines." Later, problems concerning STS management continued to be dealt with by philosophers and cybernetics experts (sometimes together with "sectoral" experts) [100]—see the brief review of *management philosophy* in [99].

Problems concerning STS management began to be discussed particularly intensively at the beginning of the 20th century, because STSs developed and became complicated and massive. The key factor is that they became massive (!), because, throughout the entire development of humankind, there have always been individuals who "worked," "balanced," and explored at the edge of a recognized complexity (at that historical time), for example, the managers overseeing the construction of the pyramids in ancient Egypt or the Great Wall of China. But they were isolated. With the waves of industrial revolutions, first, STSs became massive and, second, not just systems, but *systems of systems*—systems comprising systems (SS, System of Systems [1])—began to take shape. The realization that SSs require special approaches came much later, only in the 1990s.

Society began to feel the need to manage STSs—their creation and application—based on scientific approaches, i.e., the need arose to move from the art of management to "mass" management. Therefore, at the beginning of the 20th century, attempts were begun to understand the general patterns of the organization of biological, organizational, and other systems [17]. In the middle of the 20th century, highly complex technical systems began to be intensively created and developed (along with organizational ones). Naturally, they were included in the STS management research topic. Then came the formulation of Norbert Wiener's *cybernetics* and Ludwig von Bertalanffy's general *theory of systems* (see the review and discussion in [100]).

However, the main thrust (mainly because of the Second World War and the Cold War that followed) quickly came to focus on technical systems and the most pressing problems that could be described by suitable mathematical models; approaches like *operations research* [150]—which, using mathematical methods, examined individual fields and applied management tasks via both technical and organizational systems—quickly emerged and were thoroughly developed, but did not solve the STS management question as a whole.

In parallel, *systems engineering*[11] was institutionalized. Applied approaches, methods, and tools for the mass (!) resolution of problems involved in creating and applying complex technical systems were developed [62, 67, 137, 144]. Within the framework of *systems engineering*, areas related to systems of systems (Systems of

---

[11] "*Systems Engineering* is an interdisciplinary approach that directs and coordinates all the technical and managerial efforts required to create a system and implement it in the face of a plurality of stakeholder needs, expectations, and constraints throughout the life cycle of the system" [67, p. 10].

Systems Engineering), systems of enterprises (Enterprise Systems Engineering) (see the historical review in [48]), etc. stood out.

At the same time, mathematical methods for researching organizational systems and STSs began to develop; the *theory of active systems* [86], the *theory of hierarchical systems* [46, 87], the theory of management of *organizational systems* [21, 101], and others appeared.

A particular application of activity is business,[12] so, since the beginning of the 20th century, theoretical and applied *management* has intensively developed (and since the second half of that century it has developed *very* intensively), and in the second half of the 20th century the *theory of business processes* likewise advanced (see Sect. 2.10).

Thus, in a broad sense, the STS management problem is to some extent currently being examined in various fields of knowledge: the theories of management of organizational systems, firms, organizations, projects, etc. [101], management [88], organization theory [17], methodology [97], cybernetics [100], automatic control theory [20, 35, 142], and many other areas of modern fundamental and applied science, each of which involves dozens or hundreds of various approaches and methods.

As noted above, activity—first and foremost complex ones—constitutes the source and basis of the existence of humankind and its development. As a result, it is not surprising that activity and its elements are studied to varying degrees in various fields of knowledge (see Table 2.4 in Sect. 2.10). However, existing theories focus mainly on the elements of complex activity; for example, a great number of results have been obtained from the theory of organizational systems, and many tools have been developed to describe business processes. In spite of all this, a unified theory of complex activity, which views activity as a single complex system, has not yet been created.

Separate, known attempts have been made to create such unified models and to develop integrated approaches to the description of various aspects of activity (see, for example, [36]). For instance, the classical study by M. Porter [109] is based on a representation of the firm as a collection of various types of activity that make up a chain to create value (Value Chain). In the system engineering section, devoted to enterprises (Enterprise Systems Engineering [48, 116, 137, 144]), etc. However, none of these approaches has yet culminated in the creation of complete and recognized theories of complex activity.

The development of a unified theory of CA in the form of sets of assertions and of general and universal models is the subject of this book, which continues the *methodological direction* presented in the work by Novikov and Novikov [97–99] and their colleagues. On the other hand, complex activity is viewed as a complex system, so the research relies on and develops approaches and methods for *systems theory* (Systems of Science) and its applied expansion of e*ngineering systems* (Systems Engineering) [62, 67]. The following are used as *research methods* (specification

---

[12] *Business*—*entrepreneurial activity; work, trade, which is the source of income.*

tools): methodology instruments, systems engineering, system analysis, and facilities and tools in knowledge areas related to the analysis and description of business processes (i.e., "business process theory") and project management, which includes both theoretical foundations and Hoare's theory of communicating sequential processes [55, 56], Milner's calculus of communicating systems [90], the algebra of interacting processes [13], the Petri net theory, network calendar planning and control, discrete-event systems, partial order theory, temporal logic, synchronously and asynchronously interacting automation, process theory, system standards for quality management and quality assurance (ISO 9000: 2000 standards system [64]), etc.

Organization theory and methodology. The *organization theory* explores the phenomenological (how the system is organized), the explanatory (why this is the system is organized) and the normative (how the system should be organized) aspects of organizing systems of different nature [17, 100]. One of the subjects of the organization may be activity.

The *methodology* [97] deals with the study of general patterns of organization of activity, i.e. the methodology is part of the theory of organization (examines activity as an object of organization). The methodology includes a *general methodology* (the methodology of elementary activity), the methodology of various types of human activity (private "methodologies", to date, the methodology of scientific, practical (including management methodology as a variety of practical activity), educational, artistic and gaming activity [97]). *Methodology of complex activity* (CA) generalizes general and particular methodologies to cases of activity with a non-trivial internal structure, with multiple and/or changes subjects and technology, the role of the subject of activity in the context of his target.

The relationship between organization theory, the methodology of complex activity and the methodology of elementary activity (the general methodology [97]) is presented in Fig. 1.5, i.e. this work can be considered as an "extension" of the general methodology in the direction of the theory of organization. The most obvious "steps" that are made below in the development of methodology are that:

- in the standard description of activity, its subject and subject matter are explicitly introduced;

Fig. 1.5 Organization Theory and Methodology

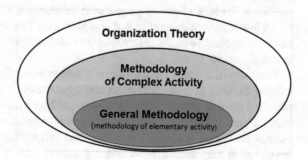

- a model of the structural element of activity and the rules of its operation for constructing a fractal structuring of activity are proposed;
- constructively describe the logical, cause-effect and process structures of activity, reflecting, including the structure and content of the implementation phase of the activity;
- key roles of technology and uncertainty of activity are identified and analyzed;
- a joint coordinated review of the life cycles of activity, its subject, subject, resources and technologies;
- the organization and management of activity are analyzed.

All this allows us not only to investigate the general patterns of complex activity, but also to solve the problems of organizing and managing it, including its subject and/or subject to complex organizational and technical systems.

Indicators of the relevance of the problems of MCA. The urgency of creating a methodology for complex activity that would become one of the general constructive grounds for the management theory of OTC, on the one hand, and their absence, on the other hand, is illustrated by specifying several significant indicators.

The first indicator is as follows. To date, a large number of methods, approaches, tools in the fields of knowledge and practice related to the management of OTC have been developed and continue to be developed. For example, one of the online sources [60] contains a catalog of methods and approaches in the field of *management*, which includes more than two hundred (!) directions, among which are dozens of well-known and proven in practice (for example, business process reengineering [53], "philosophy Kaizen" [82] or management by the goals of P. Drucker, etc.), and little-spread.

However, the presence of a large number of similar methods and approaches does not provide a holistic view of the functioning and management of organizational and technical systems (see also Table 2.4). There is no universal formal tool to justify the optimality of proposed and/or adopted decisions, audit of management systems.

A generally accepted approach to training managers is the "method of *case study* methodology, using a description of real economic, social and business situations." The case method was introduced into practice by professors at the Harvard Business School and is now widely used by universities around the world [54, 85]. The prevalence of the approach based on the analysis of concrete examples ("cases") testifies to the lack of an adequate theory of management of this class of systems, which has sufficient generality and completeness. This approach, of course, is also limitedly applicable to the creation of new management and management systems in new situations.

Generally speaking, all traditional management relies on heuristics: with the existing number of diverse theories and approaches, it is impossible to determine which of them is effective and in what cases? do I have to practice everything at once? or one? which the? or some set? and how to form this set? In management, there is no formal basis for comparing and optimizing management systems. So, strange as it may sound, most of the practical activity is managed heuristically, and, consequently, it is not optimally realized. Solving such issues requires a single constructive foundation, and this work is an attempt to develop and present this basis.

The second indicator is the trend in the last decades of a significant transformation of the principles of linkage formation in organizational and technical systems, primarily the reduction of the role of organizational links, organizational structures, the manifestation of interest in systems of very different nature with a weak or completely absent rigid structure. This trend has three types of manifestations.

First, the modern development of such new technologies for the organization and management of production as networked, *extended* and *virtual enterprises*, global production and production as a service, Internet of Things, Industry 4.0 Concepts [126] or Smart Manufacturing require deeper and broader integration and interoperability of activity—it's not enough to connect sensors, actuators and controllers, integrate automatic control systems with shop floor and enterprise level management systems. *Integration* has become global, a special form of organization such as expanded enterprises—aggregates of enterprises and firms, united by unified technological processes and connections without legal and financial association, has developed in a special way. In the expanded enterprises, the main are the technological links, and not the organizational structure or the share capital.

Secondly, in the sphere of management, the idea of changing organizational paradigms, replacing managerial rigid structures with platforms and *"functional houses"* (pools of homogeneous resources), increasing the value of flexibility and the speed of response to changing conditions, are becoming increasingly popular [32]. All this means the transfer of managerial ties from rigid organizational structures to operational *technological ties.*

Thirdly, theories and approaches of "controlled chaos" [78], "complex Adaptive Systems" [51, 59] are developed and applied in geopolitics and similar spheres of analysis and management of social "mega-systems" or systems consisting of systems [47, 96], various network models [94] and other similar systems that consider systems (mostly social) in which the internal structure disappears or is practically absent. In computer science and in artificial intelligence, the number of papers devoted to formal means of modeling such systems, for example, *multi-agent systems* (see reviews and discussion in [38, 59, 120, 128, 155]) is increasing.

All this allows us to fix the change in the role of ties, the weakening of structures, the manifestation of global factors and the integration of activity. That is, the structures of the systems are "blurred", disappear, but the systems continue to function; this means that the elements of the TSS carry out coordinated activity with transforming ties between them. Under such conditions, shifting the focus of the research to the analysis of the *activity* of the elements can be very productive, which forces the development of unified approaches and models of complex activity.

The third indicator is that the explosive growth of data volumes and the ability to accumulate and process them created new challenges for classical fundamental science. Approaches related to the areas of *artificial intelligence* [133] Big Data or Cognitive Sciences, have become very popular in recent years, which is due to their high efficiency in approximating decision functions based on the available historical data. But for practical application it is necessary to be able to explain the admissibility of using the formed decision functions: to prove monotony, stationarity, the stability of the phenomena being studied, to confirm the legitimacy of using approximations for making decisions based on future new data. For their successful application, a tool is needed that allows:

- reasonably combine large amounts of current data and the results of their processing with a priori representations of expert experts and with fundamental laws;
- interpret the data and the results of their processing;
- generate and select useful hypotheses.

Successful development of these approaches in a new way raises the problem of organizing the interaction of natural intelligence with the artificial, the problem of the joint work of the machine and man (see, for example, [11]). Actually, the tasks of creating organizational and technical systems of a new quality, realizing the integration of human activity and the "activity" of the machine, included in *augmented reality*.

The fourth indicator is a certain stagnation in the *psychological theory of activity* (Activity Theory), which developed intensively in the middle of the 20th century and became the universally recognized basis for many modern sections of the STS management theory—see the reviews in [21, 101].

Indeed, in the framework of the psychological theory of activity (Rubinshtein and Leontief [76, 119])—see reviews in [39, 40], considering the activity of a person or groups, the activity has a complex three-level structure in which it is possible to distinguish a hierarchical structure "activity → action → operation" and psychophysiological functions (activity is determined by motive, action goal, operation by specific conditions of its realization).

The main foreign follower of the school of the famous Soviet psychologist A. N. Leontiev is a Finnish psychologist and educator Yu. Engstrom [41]; his first monograph [39] was published in 1987, the second edition [40]—in 2012. Most of the modern research on the theory of activity refers to its non-hierarchical (!) Structure of activity [39, 78] (mainly educational activity)—see, for example, [114].

To date, the theory of activity is actively used and is developing, probably only in the models of human-computer interaction (see, for example, the special edition of the Scandinavian Journal of Information Systems (2000, Vol. 12)).

Here, perhaps, and all that can be briefly said about the current state of the theory of activity. It is unlikely that the activity of a complex STS can be described by a three-level A. N. Leontief scheme or the "network" structure of J. Engestrom. Therefore, it is urgent to develop multi-level hierarchical models of CA, allowing describing its complex structure in a uniform manner, to operate with its elements, including in their dynamics and development.

All this makes the problem of creating and developing a methodology for complex activity urgent.

1.3 General Research Outline

The present study follows a unified scheme, which, together with the structure of the presentation of the material, is presented in Fig. 1.6.

First of all, the authors, based on their professional experience, analyzed numerous examples of management of organizational and technical systems in such different forms as the firm/enterprise/organization, their subdivision, working group, project, national project or program, municipality, state agency, regional or national economic association, etc. Consideration of the examples led to the conclusion that the subject of management should be not only and not so much the STS itself, but, above all, the realization of *complex activity* (CA) by them, i.e., the activity in which they are the subjects.

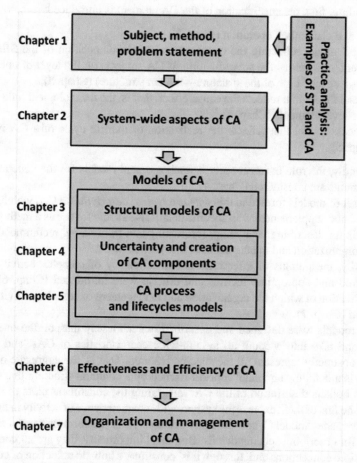

Fig. 1.6 The methodological scheme of the study of CA and the structure of presentation

Based on the results of the analysis of examples of management of STS and generalization of CA characteristics, general aspects of CAs characteristic of any CA were formulated. Since the CA is a complex system, these aspects are called system-wide, emphasizing this name, that they are the general basis for distinguishing the class of "complex activity" systems and considering different instances/examples of CA as single-species entities. Unlike system-wide, other characteristics or elements of specific realizations of CA are conditionally called *specific*.

Next, an integrated set of CA models was synthesized that reflect the system-wide properties of CAs, describe it as a system in the unity of its components (needs reflecting, among other things, demand, goals, technologies, etc.) together with subjects and objects [8, 10] on the whole *life cycle* of the CA (the time interval of its existence from the moment of origin/fixing of the demand to the completion of the CA).

Including, first, the specification of the CA element is introduced:

- SEA and elementary operation (Chap. 3);
- Logical model reflecting the composition and decomposition of the SEA into elements, as well as the subordination of CA subjects (at the level of one SEA, this is a special case of the structure—the fan structure) (Chap. 3);
- a causal model that reflects the cause-effect, that is, the technological links of the lower-level elements (Chap. 5);
- a process model that reflects the realization of the life cycle of a CA in time (Chap. 5).

Secondly, the role of uncertainty is analyzed, and models for the generation of CA elements are considered (Chap. 4).

The set of models formed in this way is a *typical description*, a CA, which, when specified and supplemented with specific details, is suitable for use as a methodological basis for describing, analyzing, forecasting and formulating recommendations for the organization and management of any particular CA.

Thirdly, the notions of effectiveness and efficiency of complex activity were considered and appropriate metrics and criteria were introduced (Chap. 6), with the application of which the organization and management of complex activity were analyzed (Chap. 7).

The models were designed and described in such a way that, on the one hand, reflect and take into account all system-wide characteristics of CAs, and on the other—abstractly represent all the specific features. That is, the aggregate of these models is a definite "template" a priori formed by a common structure that creates a single holistic description of the CA, integrating the constituent parts of the two types. The first of them express the system-wide characteristics of activity in the form of ready-made "models", and the latter are created in the form of certain abstract "slots" for describing specific details. But, most importantly, they are all connected by multiple connections and, through this, constitute a holistic reflection of complex activity as a complex system.

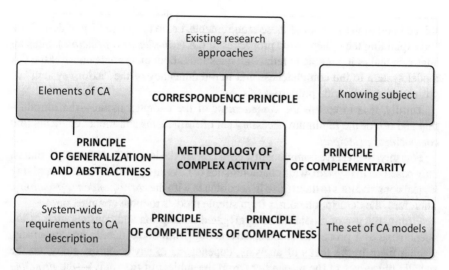

Fig. 1.7 The logic of highlighting the principles of studying CA

Several general principles were laid down in the basis of this study (see Fig. 1.7). The first two are a consequence of the fact that complex activity is a complex system. And the third, fourth and fifth reflect general approaches to the study of CA.

First, in the course of the study, it was necessary to review the CA from various points of view, to involve a number of existing research approaches in several areas of knowledge (see above and Sect. 2.10), primarily methodology, systems theory and system engineering, and the result of research was expressed in a set of interrelated models (not contradicting and developing known results—the *correspondence principle*). The plurality of views (including cognitive subjects) and models reflects, on the one hand, the *principle of complementarity*, and on the other, the system-architectural approach, which is one of the foundations of system engineering (for more details on the system-architecture approach, see Sect. 2.10 and [68]).

Secondly, when constructing any theory, the question arises of its *completeness*; in this case it is a question of the completeness of the descriptions obtained in terms of the developed models. Complex activity, being a complex system, has a number of features characteristic for such systems (Chap. 2 is devoted to an analysis of the features of a CA).

Due to the multiplicity of links, structural fractality and, especially, the uncertainty of CAs and the generation of new elements of activity in the CA process, the question of completeness of any description becomes extremely difficult. So, in the presence of any predefined set of CA models, in view of its marked features, it is always possible to offer one more model, which also describes the essential properties of the CA, but different from the available models. Therefore, within the framework of the study, the system-wide properties of complex activity are singled out (Chap. 2). They are fixed as requirements to the theory being created (Sect. 2.10), and the question of the completeness of the theory being developed—the system of models is put and solved in relation to the set of these requirements, i.e. to the possibility of describing and explaining the system-wide properties of CA (*completeness principle*). Such an approach makes it possible to reduce all questions about the completeness of the CA model system to the completeness and non-redundancy of the "axiom system" of postulated system-wide properties of CAs.

Thirdly, it is to ensure the *compactness* of the complex model—the introduction and use of the minimum necessary set of formalisms and model elements (the *compactness principle*).

Fourthly, to preserve commonality, each new model is built "consistently", that is, only based on the system-wide characteristics of CAs and previous models, applying logical constructions to them (also in accordance with the *correspondence principle*). Therefore, the description comes from simple models to more complex ones.

Fifthly, the use of a detailed or aggregated description and consideration of the phenomena and processes studied, abstracting from some details or, conversely, placing them in the focus of analysis, depends, as in any scientific study, on the specific objectives of the researcher (from the subject of the study)—the *principle of generalization and abstraction*. Due to the specifics of this work, this approach is especially often applied to the elements of activity: practically every element of activity in one case can be adequately represented as elementary and unstructured, and in the other it can and should be considered as complex and having a complex internal structure.

Summary of This Chapter

The subject of the present study is defined: this subject is complex and includes a complex interacting and interrelated pair "complex activity and organizational and technical system". CA is the primary subject, and OTC is mediated, acting in the roles of the CA subject and/or its subject.

The definitions of management and organization as special cases of complex activity are introduced and considered:

Management is a complex activity that ensures the effect of a management subject (the subject of this CA) on a managed system (management entity) designed to ensure its (its) behavior leading to the achievement of the objectives of the management entity.

Organization is a complex activity with the aim of creating internal orderliness, coherence of interaction between more or less differentiated and autonomous elements of the subject of this activity (including through the formation and maintenance of interrelationships between the elements with given characteristics).

The relevance of research into the management of CA and STS is illustrated by four important indicators:

1. Existing approaches and the results of modern methodology and psychological theory of activity do not allow describing the activity of modern complex OTCs in a uniform manner, for which it is necessary to develop an adequate formalism that reflects, among other things, the complex hierarchical nature of the goals and results of the CA.

2. The existence of a large number of developed methods and approaches in various fields of knowledge does not provide a holistic view on the management of CA and STS, there is no universal formal tool to justify the optimality of proposed and/or adopted decisions.
3. The trend in recent decades to reduce the role of organizational ties, organizational structures, the manifestation of interest in systems of very different nature with a weak or completely absent rigid structure. That is, the structures of the systems are "blurred", disappear, but the systems continue to function; this means that the elements of the STS carry out coordinated types of activity with weaker links between them.
4. The explosive growth in the volume of data and the ability to accumulate and process them has created new challenges for classical fundamental science, which in a new way poses the problem of organizing the interaction of natural intelligence with the artificial, the problem of the joint work of man and machine.

A general scheme for constructing the MCA is presented in Fig. 1.6. With reference to the study of complex activity, principles are formulated: the principle of complementarity, the principle of completeness, the principle of compactness, the correspondence principle and the principle of abstraction and generalization.

Chapter 2
Complex Activity and Its System-Wide Character

In the second chapter, the concepts of elementary and complex activity are introduced, system-wide features of complex activity are revealed, on their basis requirements to the developed theory of the methodology of complex activity are formulated.

The material of the second chapter includes ten sections. In the first section the definitions of elementary and complex activity are given, sections from the second to the ninth are devoted to the isolation and analysis of system-wide features of CA. In the tenth section, based on a brief analysis of adjacent areas of knowledge, the rationale for the creation of a methodology for complex activity is presented and requirements for it are formulated.

2.1 Simple and Complex Activity

The generally accepted representation of the procedural components of the activity in modern methodology (see Fig. 1.2) does not presuppose the decomposition of the goals, technology and result of the activity and does not contain corresponding procedures. The description of the result or technologies in the form of non-detailed and unstructured objects significantly narrows the applicability of this approach for the study of practically interesting non-trivial examples of activity.

For example, the presentation of activity to develop a model of a car or a washing machine, or any other complex technical facility, for its production, sales, service, etc. at the level of abstraction Fig. 1.2 will not allow you to obtain any results useful for analyzing the activity or creating a management system for it.

A similar example: a civilian aircraft (aircraft, aircraft) is a complex system, and as a consequence, requires a variety of sophisticated technologies for its design, production and service; consequently, and various OTC, implementing these technologies. Therefore, the representation of the production technology of the aircraft, the aircraft

M. V. Belov and D. A. Novikov, *Methodology of Complex Activity*, Studies in Systems, Decision and Control 300, https://doi.org/10.1007/978-3-030-48610-5_2

itself and the subject-everyone as a whole, does not appear to be constructive and operational.

The same conclusions can be drawn for the retail bank, its lines of business, branches and branches, as well as the shops and divisions of the nuclear power plant, the fire department and other "cross-cutting" examples considered in this paper.

Therefore, we introduce the concepts of elementary and complex activity. By *elementary activity* we mean such activity, goals, technologies and the result of which do not have their own non-trivial internal structure (or, according to the principle of generalization and abstraction, the introduction of such a structure does not provide additional knowledge, a qualitatively new effect). Characteristics of elementary activity imply that three conditions are fulfilled:

(1) its technology does not change during the activity, the activity itself is clearly defined for the observer (researcher, subject of activity, consumer of its results) framework;
(2) the subject of the activity is unique and varies in its process according to the technology (which, in fact, constitutes the goal of the activity), but does not change its place and role in the context (requirements to the subject of activity do not change);
(3) the entity performing the activity is unique, clearly defined and limited, and also not transformed in the process of carrying out the activity.

In the case of elementary activity, the structure of Fig. 1.2 is sufficient to describe the activity itself, and there is no need to consider the subject and the subject together with the actual activity—they play the role of an understandable context (during the activity period only the subject evolves in accordance with the technology used by the subject).

In contrast, an activity that is not elementary will be called complex. That is, *complex activity* is an activity possessing a non-trivial internal structure, with multiple and/or changing goals, subject, technology, the role of the subject in its target context.

The variety of elements of complex activity allows us to distinguish several characteristic groups among them. First of all, these are elements of "*core activity*", which is directly aimed at changing the subject, to achieve the ultimate goal. The other two groups form elements related to *management* and *organization* (as a types of activity). Finally, it is appropriate to refer to the fourth group the elements of "*auxiliary activity*", the goals of which are related to the procedural components of other elements of activity, primarily with technology and resources.

Obviously, the introduced typology is relatively conditional, because in each specific case there will exist elements that can be assigned to different classes according to various characteristics (in particular, all elements of the CA are referred to as "main activity", since, ultimately, they provide the achievement of its goals). However, such a grouping is advisable, because on the basis of the principle of generalization and abstraction, each element of the CA can be constructively described and studied within the framework of the characteristics of a particular group. So, along with models describing any elements of the CA (Chaps. 3 and 4, Sects. 5.1 and 5.2), models of the elements of "auxiliary activity" (Sects. 5.3 and 5.4) are presented below, and a separate chapter (the seventh) is devoted to management and organization.

2.2 Complex Activity Structures

Logical structure of CA Complex activity is characterized by multiple technologies, subjects and/or objects of activity, when elements of one complex activity are elements of another complex or elementary activity. It makes sense to note the *"fractal" property* of the CA: in many cases, the CA is decomposed into elements, which in turn are CAs, the elements form multi-level hierarchies—the *logical structure* of the CA.

The higher-level elements in such hierarchies are not only decomposed in the form of lower-level collections, but also, in a certain sense, induce subordinate to perform the activity and are consumers of the results of their CA. Therefore, we will say that the higher-level elements of the CA create an incentive *demand* (or simply demand) in relation to the activity of the downstream and are *users* or *consumers* of the result of their CA. Demand can be deterministic or indeterminate, depending on this, various models of CA elements will be introduced below. The demand is actualized by the subjects of the subordinate elements and "becomes" their needs (Fig. 1.2), for which they perform activity, the results of which in turn "consume" the higher-level elements of the CA.

This process of "unfolding" the CA structure over time and its individual aspects are discussed in more detail later in this chapter and in Chaps. 3 and 4.

Within the framework of the illustrative examples considered in this paper, the following logical structures of CAs can be distinguished.

The CA of the retail bank as a whole is divided into CA offices (and/or branches) and CA departments of the central apparatus. CA offices include direct customer service CAs, control CAs, auxiliary CAs, administrative CAs, and so on. CA of direct service consists of CA of crediting, CA of conducting of deposits, CA of settlement-cash service, etc.

The CA of the aircraft assembly (one of the CS elements of the aircraft production) is structured on the CA of the individual sections of the assembly, and those in turn—on CA operations. Each of the elements of the CA structure has its own subjects (possibly coinciding in some cases)—workers or groups of workers, the objects of the CA elements are the same physical objects—aircraft instances in different stages of readiness. The lower level of decomposition of the logical structure of a CA is an elementary operation (in project management called work).

The fire department's fire department includes a fire department's fire department, an auxiliary services CA and a corresponding control system CA.

CA for generation of electricity at a nuclear power plant, one of the elements has a CA of the current content of fixed production assets. Current content includes such

CAs as routine inspections and preventive maintenance, which are implemented for various types of equipment, buildings and structures.

In each of the examples, complex activity generates organizational and technical systems, including both technologies and technological complexes, as well as providing systems, and multiple subjects, and multiple objects.

Causal (temporary) structures of CA. Elements of complex activity are in the cause-effect relationshipes with each other, forming a *temporary structure* of CA.

Indeed, lending to a particular individual (as an element of a retail bank CA) includes the following operations:

(a) receipt of an application and questionnaire from a potential borrower;
(b) collecting additional data about it;
(c) risk analysis and preparation of the borrower's internal form;
(d) consideration of the borrower's questionnaire by the relevant employee or commission and making a decision;
(e) execution of a loan agreement;
(f) making postings on accounts.

Similarly, assembly operations at each section of the final assembly of the aircraft are regulated by technological cards that determine the sequence of actions, the tools, materials used, work methods.

The actions of each member of the fire crew at the exit on call are described in the instruction and are repeatedly practiced during training to achieve the greatest efficiency.

Strict regulations determine the actions of the personnel of the nuclear power plant. For example, when the unit is scheduled to shut down for the next fuel reset (once every 12, 18, 24 months depending on the fuel cycle being implemented), planned repair and control works are performed in parallel, as well as upgrading the equipment of the primary circuit, which can not be performed while the reactor is operating. The duration of such stops is several weeks, during which dozens of works are performed, including thousands of operations in clearly defined sequences.

More details of CA structures are discussed in Sects. 3.3 and 3.4.

2.3   Life Cycles and Complex Activity

In various fields of practice—production, management of organizational and technical systems (firms, organizations, projects), the concept of life cycles (LC) has become popular nowadays [62, 67]. Under the *life cycle*, we will understand (based on the definition [67]) the evolution of the system, product, service, project or other object, from concept (or appearance) to disposal (or termination of existence).

The life cycle is usually considered as a set of *stages* (possibly parallel and overlapping with each other in time); in [67] the most common stages of a complex artificial system are distinguished: the concept, design, production/creation, application, support and utilization. The concept of the LC is also widely applied to organizations, businesses, project programs, employees, production assets, technologies, knowledge. The spread of this concept is caused by the need to more effectively manage the processes of creating and operating objects of activity. As a rule, it is necessary to bear significant expenses at some stages of the existence of an object, while its usefulness "manifests itself" on completely different ones, therefore, consideration of only certain stages can give much distorted results.

The concept of the life cycle will be used for several "objects". Firstly, the LC of the system, product or object is one of the forms of CA organization, and in this case we will speak about the LC of the *subject* of activity or the LC of the *subject matter* in the sense of the definition given in the previous paragraph.

Secondly, the CA itself, being a complex system, evolves in time (from the moment of origin or fixation of the need to the completion of actions and reflection). Therefore, we introduce the concept of the *life cycle of a complex activity* (or its element) as a complete process, including fixing demand and understanding needs, setting goals, structuring goals and tasks, selecting and developing technology, performing actions in accordance with technology, obtaining results, evaluating results and reflecting. LC CC is actually a "deployment in time" of the procedural components of the activity; its structure is shown in Fig. 2.1.

Following the established tradition in system engineering, the temporal structure of the LC will be represented in the form of stages, and at the same time it is advisable to combine the stages of the LC of the CA into phases, as is done in the methodology

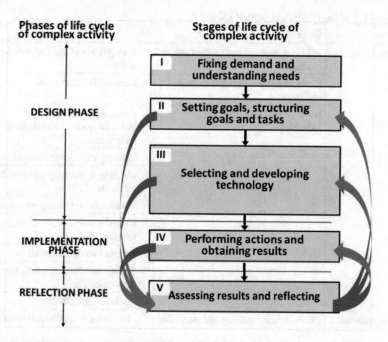

Fig. 2.1 Phases and stages of CA life cycle

[97]. In general, the process of realizing the LC of the CA is iterative; its stages can be repeated and overlapped, especially in the stages of the formation of technologies and the implementation of actions. Iteration of the LC CA reflects the reflexive nature of complex activity, one of the cycles of reflection characteristic of it.

One of the reasons for the concept of the LC is the separation or distribution of the time periods for obtaining certain effects, benefits, usefulness and application of effort and costs.

Consider life cycles as a form of organization of CA (analysis of the life cycle of complex activity and the creation of appropriate models is devoted to Chap. 5), introducing the phases, stages and stages of the CA, shown in Fig. 2.1 and in Table 2.1 (see also the more detailed discussion below).

Most clearly (among the illustrative examples used), the need to consider the life cycle of an object CA (i.e. LC as a form of organization of CA) is manifested for the fire department and nuclear power plant.

The usefulness of the fire department is realized only in short periods of direct work of firefighting in the course of extinguishing fires. The costs of the fire department itself during these periods are incommensurably small in comparison with the content, training of personnel, equipping with modern technology, the content of this equipment, etc. Therefore, consideration of the fire department only during periods of firefighting without taking into account the remaining stages (and vice versa) will be incorrect.

Table 2.1 Phases, stages, and steps of LC of CA

Phase	Stage	Step
Design	I. Fixing demand and understanding needs	1. Fixing demand and understanding needs
	II. Setting goals, structuring goals and tasks	2. Creating logical model
	III. Selecting and developing technology	3. Checking the readiness of technology and the sufficiency of resources
		4. Creating cause-effect model
		5. Creating technology of lower-level elements
		6. Forming/modernizing resources
		7. Calendar-network scheduling and resource planning
		8. Performing optimization
		9. Assigning actors and defining responsibilities
		10. Allocating resources
Implementation	IV. Performing actions and obtaining results	11. Performing actions and obtaining results
Reflection	V. Assessing results and reflecting	12. Assessing results and reflecting

The usefulness of nuclear power plants is realized during periods of generation of electricity, and during these periods only operational activity is carried out, which accounts for slightly more than 20% of the costs of electricity. The greater part of the costs price are the costs of nuclear fuel (about 30%) and the capital costs for the construction, repairs, modernization of the power unit itself (almost 50%), and these costs are realized at the stages of unit construction, modernization, shutdowns for overhauls and fuel transhipments. Therefore, separate consideration of activity at different stages will not give a complete picture.

The aircraft production activity also need to be considered in conjunction with the CA for the design of the aircraft, the production preparation CA and the follow-up CA because of the strong interrelationships between them.

The examples given above demonstrate the necessity of joint consideration of successive stages of the life cycle of an object of CA. Similarly, the need to analyze the parallel stages of the LC, this is most clearly manifested for the stages of operation (use, application for the goal) and maintenance (routine maintenance, repairs, etc.) of the item. Traditionally, work on the content of the subject (diagnostics, maintenance, repair) interrupts its operation (it is sent to aircraft repair enterprises for performance of routine maintenance of the aircraft, diagnostic and, moreover, repairs on the nuclear steam generating unit are performed with the reactor stopped, etc.). However, the development of modern technologies for monitoring and diagnosing equipment, together with information and communication technologies, leads to a wide spread of integrated diagnostic systems. These systems allow, first, to shorten the period of interruption of operation for diagnostics, and secondly, to switch to methods of servicing by condition and predictive maintenance [111]. For example, most models of modern cars are equipped with built-in computer diagnostic systems that display results on the dashboard. The widespread use of mobile Internet has led to the possibility of collecting and processing this information by diagnostic stations and insurance companies. All this significantly increases the economic characteristics of the life cycle of the CA, and from the point of view of the methodology, the CA "blurs" the boundaries between the GC stages.

The examples considered above can be generalized for the phases and stages of the LC of the CA, defined above (Table 2.2), their system-wide characteristics, choosing as the basis "obtaining effects against the required costs".

Another significant trend, which strengthens the need to consider the subject throughout its life cycle, is the expansion beyond the project (!) Period of operation of complex, mainly high-tech products and facilities due to the use of reserves laid during the design and deep modernization. This allows obtaining a significant economic effect, reducing significant capital costs. Examples of this trend are multiple extensions of the service and modernization of strategic bombers Tu-95, Tu-160 (USSR-Russia) and B-52 (USA), radical rearmament and the introduction of the battleship Iowa (USA) in 1984, which was launched in 1943 and canned in 1958. Also, the majority of nuclear power plants that were built in the 1960–1970s have developed their design life. However, due to the multiple constructive reserves of characteristics, the main equipment (nuclear steam-generating plant), buildings and structures can be safely used for many years. Therefore, IAEA experts prepared

Table 2.2 Cost-effect analysis of phases and stages of LC of CA

Phase of LC	Stage of LC	Costs	Effect
Design	Fixing demand and understanding needs	Low costs, because CA has the form of elementary operations	No effect, and no effect expected
	Setting goals, structuring goals and tasks	Costs can be high or low	
	Selecting and developing technology		
Implementation	Performing actions and obtaining results	High costs	Effect expected or gained
Reflection	Assessing results and reflecting	Needs are exhausted and CA is completed; hence, costs are low	No effect

and approved regulatory documents to extend the life of such nuclear power plants. Attempts to create reusable space systems, begun as early as the 1980s and continuing now, are also aimed at improving the economic parameters of their life cycles.

Using new construction materials can develop this trend and make it massive. Stability and non-degradation of the strength and other characteristics of such materials can make the designs, basic aggregates and equipment of complex objects and products "eternal", then many upgrades will be economically justified, the depth and multiplicity of which is difficult to predict when designing, creating a product, system or object.

It is interesting to note that this trend is not historically new: the builders and engineers of ancient and medieval temples were not asked by the "project lifetime" as a requirement for design and construction, and often by default defined the lifetime as "the object must exist forever". Multi-stage completions and modernization lasted for millennia. For example, the first building of St. Peter's Cathedral in Rome was built in the IV century as an ancient Roman basilica, a modern architectural form the cathedral received in the XVII century; separate structural bearing elements of the Cordoba Cathedral were erected during the periods when the structure was completed and served as an ancient Roman basilica, an early Christian cathedral, a mosque and again a Catholic cathedral. Also, the engineers of the first rail-based steam-powered transport systems (for example, the London Underground or the Nikolaev Railroad) could not imagine what profound modernization and development would survive the "objects of their activity" during life cycles that are not yet over.

Thus, there is a need to consider the subject of activity at all phases and stages of the LC, in all forms. It becomes necessary to analyze the activity associated with this subject, also at all its stages. In fact, these are various types of activity that are connected in a single way with a single subject and "*information model*" with information about it and about the technologies used (the role of the information model is discussed in Sect. 2.6).

In many cases, a complex activity on the whole life cycle of an object is characterized by the repetition of the elements of activity, then the judgment about the activity (its effectiveness or effectiveness) can be made on the basis of not a single case, but only a representative sample.

For example, the activity of a retail bank is not limited to servicing one or several clients within a few hours. If the analysis is the CA of the bank as a whole, its line of business (various types of banking services, for example, utility payments or lending) or branches, it must be done over a long period of time during which tens or hundreds of thousands will be serviced (if not millions) of customers. Similarly, activity should be considered for the production of a non-trivial number of aircraft copies and for the production of electricity over a long period of time, including stops for capital repairs, modernization, fuel loading, taking into account the initial costs for the design and construction of the nuclear power plant.

Therefore, it is necessary to consider CA during the life cycle of its subject matter of an extended time interval, on which subjects and objects of activity interact.

It is important to note that in the case of repeatedly repeated activity, for example, mass production, the concept of LC can be applied both to specific instances of products, and to collections of homogeneous copies—to the entire series. If in the first case they speak about the life cycle of the object of activity, then in the second case it makes sense to talk about the *life cycle of the technology of activity* or the life cycle of the need. In our examples, this is a copy of an aircraft against a specific aircraft model, or a banking service provided to a specific customer, against a banking service as a line of business.

In general, the concept of the life cycle can be applied to the elements of activity, technology, needs, as well as to subjects and subjects (individual or collective)—see Figs. 2.2, 2.3, 5.9 and 5.10. For more details on the application of the GC concept to the elements of complex activity, see Chap. 5.

During the LC of the CA, and/or the technologies and/or the subject and/or the role of the *object* (subject matter) in its target context can change (for example, changing the requirements for it). The object of the LC can significantly change its properties/status/form of existence during the LC; therefore, activity can also change its character accordingly. The LC of the object (see Fig. 2.2), as a rule (if not always), during its LC resides in different forms of incarnation.

First, it is in (a) the information form of the embodiment (the *design* stage), (b) then in the transitional form (the stage of creation, organization, *production*), (c) then in the target form (produced, created). In the target form in-1) it is not first used for its intended goal (*testing*, certification), в-2) then used for its intended goal, fulfills its objective functions (*operation*) and implements the expected utility. Stage в-2) can be interrupted in-3) Stages of current content and/or in-4) *modernization* stages. At the end of the use phase, the LC object again resides in a transitional form—(d) is *disposed* of, after which it can remain indefinitely in the information form (d)—in the form of retrospective data.

Figure 2.2, which depicts the stages of a typical LC, well illustrates the above thesis that the target utility of the LC facility is realized only during one stage of

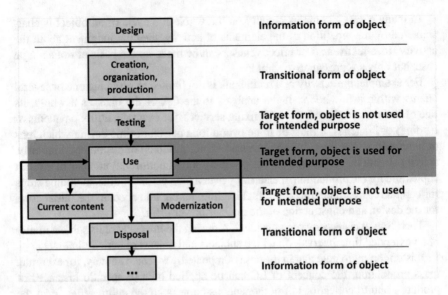

Fig. 2.2 LC stages and the object's forms of incarnation

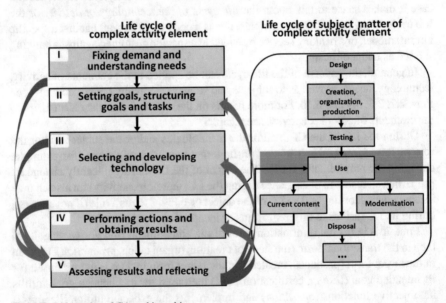

Fig. 2.3 Life cycles of CA and its subject

the intended use, while during all the others it is necessary to carry out the relevant activity and incur costs.

As noted above, complex activity is characterized by a complex structure, multiple subjects and subjects, therefore, the life cycles of complex activity (Fig. 2.1) and the life cycles of the subject of activity (Fig. 2.2) can generally be correlated in different ways. In Fig. 2.3 shows one of the options.

This option corresponds, for example, to such an element of the CA as the "Design and Production Program for a Certain Model of the Aircraft" or "Construction and Operation of a Nuclear Power Unit". Designing, preparation of production, production, modernization, operation and maintenance are stages of the aircraft engine of a certain model, which can last for decades. The modern practice of aircraft building firms is the separation of the management object and the responsibility center for the LC model; as a rule, such a center of responsibility is called an aviation program.

The phases/stages of the LC are naturally related to each other, they are united by a single external need, to which they are directed, so they naturally constitute (very, very complex) CA element. In this case, the subject of CA ("technology design and production model VS" and "power unit", respectively) exists during only one of the stages of the LC CA, which is reflected in Fig. 2.3.

In Sect. 2.2 the fractal property of CAs was mentioned, the life cycles of CA elements are also characterized by fractality—each of the CA phases can (if necessary) be regarded as an "independent" activity (complex or elementary) and/or operation with the object of activity, see Fig. 2.4.

"Technology of design and production of the VS model" and "power unit" at different stages of their development are objects of a variety of CA elements and exist much longer than the LCs of such elements (see Fig. 2.4). For example, during the creation of "technology for the design and production of the aircraft model," new workshops or even plants can be built, and new technological systems can be developed during the "power unit design".

It was noted above that in a complex CA structure the technology and the subject of one element of activity can be objects of other elements of the CA, therefore, they are objects of the LC. In its turn, the element of activity that is organized in the form of such a center (a technology or a business entity in this case) also has its own life cycle (see Fig. 2.4).

It can also be said that complex activity is realized in the form of a combination of several forms presented in Fig. 2.5.

They are listed in the historical order of appearance in practice and in order of increasing complexity ("nesting one into another"):

- Elementary operations (works) are in a certain sense atomic elements;
- Complex operations consist of elementary and complex operations;
- Projects consist of complex, elementary operations and (under) projects;
- Project programs are formed from projects, integrated and elementary operations;
- Life cycles are considered as a connected aggregate of project programs and projects.

The causal structure of CA and the life cycle of CA subject mater It makes sense to make a comment about why it is necessary to consider the *life cycle of an object (subject matter) of complex activity* separately from its cause-effect structure, although the LC is certainly a causal construction and, therefore, a particular case of

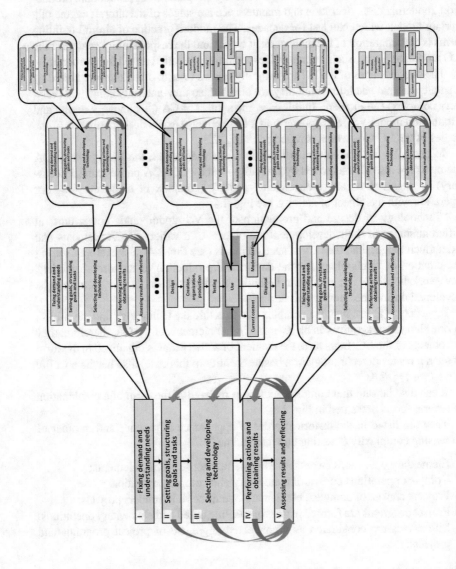

Fig. 2.4 Fractal structure of CA

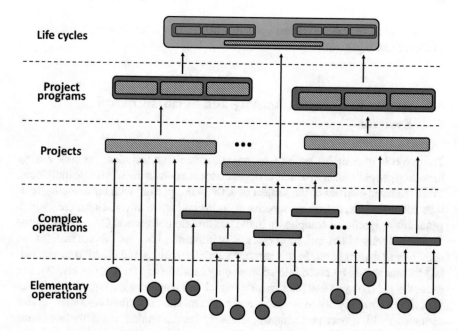

Fig. 2.5 Forms of CA organization

such a structure (Fig. 2.2 represents its diagram). The stages of the LC are, on the one hand, elements of the CA with respect to the life cycle as a whole, on the other, the stages, in turn, are decomposed into CA elements of the next level of detail.

First, in all the above examples of cause-effect structures, a certain equivalence of elements is implied. In contrast, the LC stages are substantially unequal from the point of view of the form of existence of the object, the carrying of costs and the obtaining of the target utility (see Fig. 2.2).

Secondly, if the stages of the LC are decomposed into elements of the next level, the unknown multiplicity/repeatability of the CA elements can arise a priori. For example, the number of customers served during the whole period of the branch office operation, or the number of clients to whom loans will be granted, can not be determined a priori. As well as the number of fire exits on calls and the number of aircraft produced during the production stage.

Thirdly, there is often a priori an unknown duration of the LC stages, when only the duration boundaries can be specified. For example, with regard to the serial production of aircraft, the operation of the bank's branches and offices, and fire departments, units, it is impossible (and not required) to specify the deadlines for the completion of the stage-termination of activity. Even with regard to the power units of nuclear power plants, this uncertainty has manifested itself: despite the well-defined design and construction period of operation (40, 50, 60 years), when this period is reached, many stations are working on extending the service life for several decades.

Thus, it can be concluded that special consideration of the life cycle category as a form of CA organization is necessary.

2.4 Actors in Complex Activity and Socio-Technical Systems

The subjects of complex activity are organizational and technical systems, always including people, and in a particular degenerate case consisting of a single individual. Thus, the main element of the subject of activity is a person, with his thoughts, feelings and emotions, or/and a collective with his collective unconscious, etc., which generates a significant complexity in the construction of general CA models. In the formal analysis of CA and its subject, it is important how the subject realizes the activity and does it at all. In this sense, feelings, emotions and other "transcendental" factors should be taken into account only in SEAs of carrying out activity. For example, the thoughts and feelings of a retail bank employee are completely irrelevant if he shows loyalty to the bank and performs the prescribed customer service operations with proper performance, similarly for the worker assembly line or the aircraft manufacturer's engineer, as well as for other examples of CA. At the same time, the behavior of people, of course, differs by certain purposefulness, unpredictability, etc.; therefore in this work the particular features of individual behavior considered are taken into account through the uncertainty of the CA subject (see Sect. 2.7).

For complex activity, especially for cases of its organization in the form of a life cycle, a "meta-subject" arises, which is responsible for the subject of the CA throughout its life, its entire LC. This *"meta-subject"* is "meta-STS", which carries out activity during the entire LC of the subject. "Meta-subject" is actually several STSs replacing each other at different stages of the LC.

The definition of "meta-subject" is answered by the concept of "extended enterprise" (mentioned above), which appeared in the last decades in the field of production activity management. Given the prevalence of this concept and its importance, we will dwell on it and the reasons for its occurrence.

One of the views on the economy or the ways of managing economic or production agents corresponds to the relations of ownership, the other to financial flows.

However, the main role for production is played by "through" production (technological) processes, it is these links that form modern cooperations on the basis of life cycles, it is through these processes that production itself is actually carried out.

In the past, the production results (products, facilities, services, systems) were relatively simple, which allowed them to be fully produced within the framework of one enterprise: raw materials and standard components were purchased, the finished product was manufactured and sold, and the enterprise did not deal with the aftersales service. In this situation, inter-firm technological and information relations were insignificant, joint-stock and financial management was quite adequate for the successful development of production. Other connections could be ignored "full-cycle plants" were self-sufficient.

In recent decades, products, services and the entire economy have become significantly more complicated—one plant can not completely produce a complex product; it was required to create *technological cooperation*. On the other hand, there was a need to accompany the product throughout its life cycle—for a long time, sometimes for decades.

All this has led to an increase in the role of technological and information links between enterprises. The main role was played by "through" production processes based on the LC; these processes bind enterprises with non-directive, "soft" management ties.

As a result, the most advanced firms began to take into account information and technological links associated with the life cycle of a product or service, and build a management based on such links (see, for example, [15]). There was a technology of life cycle management—grouping of enterprises on the principle of participation in the provision of the LC of the subject and management of information (primarily) and other flows within such a group. Thus, there were *"expanded enterprises"* of the aggregate of production units associated with CC products through processes and a single information space. The most adequate metaphor for representing an expanded enterprise is the "cloud" or "star" around the parent enterprise (that is, around the complex structure of complex activity), and not the hierarchy (of course, initially such "clouds" developed inside firms, between individual units, and already later—between enterprises).

For enlarged enterprises, the role of joint-stock, legal and financial ties are significantly reduced, and the role of technological ties is significantly enhanced. However, significant technological and information connections did not replace the financial and joint-stock management, but complemented it, creating another "dimension" of enterprise management.

Because of the multiple links of the organizations included in the expanded enterprises (one organization may include its parts in several extended enterprises), the boundaries of complex activity are fuzzy. For example, the activity of international companies such as Uber, or Apple, or BP, or Boeing cover virtually the entire global economic system—through users and contractors-suppliers. There are no clear criteria that would make it possible to unequivocally determine whether the organization belongs or does not belong to this extended enterprise. Therefore, in practice, the subject of the expanded enterprise decides expertly which objects to include in its control zone, and which ones are considered external interacting systems.

On the other hand, in many cases the same individuals, groups, project teams, firms and their divisions are simultaneously subjects of several CAs (CA elements). For example, employees of the bank branch as a unit, and many of them individually implement various types of CA, according to the types of banking services. Many engineers of the aircraft construction company perform tasks on several projects in parallel, the heads of engineering departments and chief designers almost always combine the management of the unit with the solution of complex engineering problems whose complexity exceeds the qualification of ordinary employees. Most of the divisions of the aircraft construction company are also involved in several aviation programs. Firms supplying aviation systems or aggregates, electrical, electronic equipment, usually work with several head firms, participating in their programs.

Thus, it can be concluded that on the set of STS as the subjects of complex activity, complex multilateral relationships are formed when several STSs are subjects of one element of the CA, and at the same time one STS is the subject of several elements of different CA.

2.5 Process and Project Activity

Consideration of practical examples of CA shows that CA elements can be conditionally assigned to one of two types; let's call them process and design types.

Process is the activity during the change of the operator of the NPP unit or the station as a whole, the activity of the consultant of the bank branch or the credit inspector, the activity of the assembly worker on the final assembly line of the aircraft. The objectives of their activity are formulated as monitoring the values of the parameters of the unit and maintaining them in specified ranges; service of all incoming customers; the qualitative performance of the assigned operations over the assembled aircraft assemblies. Goals are set, and their achievement is checked for each specific time interval (work shift or period), while the nature of goals does not change from one interval to another (target values of the parameters can change). The results of the activity can be measured quantitatively, in our examples this is the amount of generated electricity or the proportion of time during which the unit functioned in the regular mode, the number of serviced customers, the number of assembled nodes.

The main difference between the elements of the *project* type is that the achievement of goals and the results of the activity in this case are often evaluated binary: the result or obtained (coincides with the planned one or deviates from the latter), or not, the goal or achieved, or not. Partial achievement of the goal is equivalent to the lack of attempts to achieve it; the "partial" result in many cases has no objective utility. Examples of such activity include the repair team performing repair work at the power unit, designing an engine or unit of the aircraft by an engineer, and extinguishing a fire by calculation.

Obviously, process activity can be considered as a project activity if the goal is formulated as ensuring that parameters are located in the target range for a given time interval. However, for process activity, the partially achieved result has a non-zero utility: generation of electricity in volumes less than planned; customer service; assembly less than the planned number of units have value.

It should also be noted that in many cases, the process elements of the CA are decomposed into project elements and vice versa. In the examples considered, customer service (or assembly operations) during the work shift is decomposed into work with each client (assembly of each unit), which makes sense to be considered as design elements of the CA.

2.6 Complex Activity, Information and Knowledge

The realization of complex activity is accompanied by the formation and modification of the *information model* (IM) of the subject and CA. In recent decades, the importance of the information model, created in parallel with the realization of CA and the evolution of the subject, has increased significantly; IM is significantly complicated, the tasks of developing and effective application of procedures for operating information models and *"knowledge management"* become more urgent.

In this paper, the following definition of IM is used: the model of the object, represented in the form of information describing the parameters and variables of the object that are essential for this consideration, the connections between them, the inputs and outputs of the object and allowing, by submitting information on changes in input quantities to the model.

It makes sense to note that the IM existed and was always used: for example, even in Ancient Egypt, the IM temples were first formed "in the heads" of architects, then they appeared on primitive from our point of view blueprints and only then were re-embodied in the form of structures preserved for many millennia. However, in recent decades the role of IM has changed significantly:

1. For a long time, the costs and timing of the reorganization of the IM, that is, the creation of products directly, were incommensurate with the costs and timing of the development of IM, now this ratio has changed. Accordingly, the share of the costs of raw materials and materials in the costs of the finished product has decreased, and the share of the costs of development has increased. The costs of metal and composites, of which a modern car or an aircraft is made, does not exceed tens of percent in the price of the product, the rest is the costs of production of parts, final assembly and, in fact, design. The development of a new model aircraft lasts 5–7 or more years, hundreds of engineers are involved in the work, and the result is a non-real IM. A vivid example in this case is the costs of the iPhone: the costs of all the components of the iPhone 5 s (16 GB) is about 191 dollars, another 8 dollars is spent on assembling devices, i.e. the total amount is 199 dollars, and the iPhone itself was sold in 2014 in the US for $ 649. Those. a greater share in the price, along with the value of the brand, takes design and promotion to the market—that is, actually IM.
2. The information model, unlike real product, exists at all stages of its life cycle, from concept to disposal.
3. Previously, IM was created in a single development center, and production could be implemented in the cooperation of contractor manufacturers. Now the process of product development, that is, the creation of IM, is carried out in branched cooperation [79]. For example, it is known that the development of Boeing CA aircraft is carried out by several engineering centers in the US, Australia and Russia, similarly—Airbus.
4. In the past, information models in the form of drawings, specifications and other technical documentation were used only for production, now it is not. For example, practically from the very beginning of nuclear power development,

the safety requirements for power units are confirmed on the basis of calculations, that is, on the basis of IM. In recent years, IM has also been used to certify auto-mobile, aircraft and other technical facilities.

5. With the complication of production facilities and the expansion of the use of IM, the models themselves are becoming increasingly complex and expensive. Now the IMs contain not only a geometric description and structure of the product, materials and technological maps, logistical information, but also complex models of functioning, movements, and others. IM is a complex hierarchical multidisciplinary complex.

6. The process of separation of the "material" and "intellectual" parts of production is becoming ever more intense. Formation of the "intellectual part" of production as a self-sustainable sub-sector—creation, modernization, etc. takes place. information models. In addition to the growth in the number of companies whose business is engineering services (this was previously), new factors are emerging: first, the standardization of engineering services, and as a consequence, "offshore developments," and, secondly, a significant increase in the costs of the "intellectual part "Production/product in relation to" material ", primarily because of the complication of products of production. The economic shift of emphasis from material objects to intellectuals is supported by the trend of separating the "intellectual part" into an independent sub-sector.

7. "Intellectual part" of the product of production—the information model is also institutionalized and increasingly becomes a commodity. The complete IM includes information components and marketing, and requirements, and the results of design/construction, and manufacturing technology, and certification, and logistics, and plans, and the structure of the cooperation of the manufacturer, and other attributes. IM is formed, changed and used at all stages of the LC by all participants in the cooperation. The role of IM is enhanced—it is used not only for the manufacture of a product (for example, as a basis for the design and technological organization of production), but also for the certification of products and other purposes.

In fact, complex activity in the field of complex STSs have evolved into two parallel and interrelated processes:

(1) creation and maintenance (including modification) of the information model;
(2) the implementation of actions (impacts on the object) in accordance with this model, ensuring evolution during its life cycle—that is, actually the activity.

This transformation was an objective source of the revision of the role of information in the life of society, manifested in numerous discussions of "information explosions", "transition to an information society", "digital economy", "knowledge economy", etc.

The information model generally contains not only normative, a priori information on complex activity, but also operational (in relation to specific objects) and forecast, as well as various historical historical data, auxiliary information with a different level of detail and formalization.

Complicating IM and increasing its role objectively causes the need to create effective methods and tools for its creation, storage, use, modification, maintenance of integrity and so on. It is these methods, procedures and tools that are the subject of several areas of knowledge and activity that is a part of the broad "information technology" industry. The popular and widely discussed technologies for product lifecycle management (PLM) and knowledge management are of great importance for the creation, maintenance and use of complex IT activity, see, for example, [30, 31, 37].

Product lifecycle management is defined [28] as a strategic business approach for applying a consistent set of tools that support the joint creation, management, dissemination and use of product information within an extended enterprise, from the product concept to the end of the life cycle, integrating employees, technological processes, production systems and information. The methods and implementations of PLM-class software are very widespread: virtually all modern production activity is based on their use. According to leading analytical agencies, the market for PLM products is tens of billions of dollars per year [29] and continues to grow rapidly.

Knowledge management is designed to operate with a wider and less formalized range of information. It is defined in [37] as a discipline, developing an integrated approach to the definition, collection, systematization, search and provision of all enterprise information assets, which may include databases, documents, policies, procedures, as well as the knowledge and experience of employees not yet collected. The main tasks of knowledge management are the classification and structuring of professional knowledge, the creation of databases and repositories of professional experience and expertise, the organization of professional communities and information exchange within them.

It can be provisionally said that if PLM tools and methods provide management of the IM of a CA subject, then knowledge management is the management of the information model of the CA as a whole.

2.7 Uncertainty in Complex Activity

In general, uncertainty is characteristic for any activity, and it is of considerable importance, especially for CA. When considering the uncertainty of the CA, we will use the ideas of Knight [70], who, first, shared measurable and true uncertainty, and secondly, considered uncertainty a source of not only negative problems, but also development.

The *uncertainty of the CA* will be referred to as the possibility of the occurrence, during the CA, of certain events that affect the realization of the CA and its outcome, but which may or may not occur. The consequence of the uncertainty of the CA is the inability to predict a priori the characteristics of the result of the activity, the moment of its receipt and the efforts (resources) that will be spent for this purpose.

The *measurable uncertainty* of a CA is defined as the possibility of occurrence of events described by certain laws (which may or may not occur). For the analysis of such events, quantitative methods (for example, probabilistic/statistical) based on previous measurements or fundamental laws (together with the assumption of the invariance of conditions and regularities) can be used.

The *true uncertainty* of the CA is the possibility of the occurrence of unique (or rarely repeated) events, which are not explained by the existing fundamental laws, and for which there are no a priori observations. In project management, true uncertainty is sometimes called unforeseen risks.

The principal difference of true uncertainty from the measurable one is that the events of the first of them arise from unknown factors (frequent and important, but a special case—the active choice of the individual), while the events of the second, although unpredictable, are described by certain regularities.

The subject of analysis and attempts to comprehend and predict unique events, their emergence and impact on complex (primarily, social) systems has become popular in recent decades (for example, the "theory of black swans" [140] and "anti-fragility" by Taleb [139], catastrophe theory [112]). This is explained by the fact that it is these events that lead to changes in the models of the development of systems as a whole-a change in the tendencies in the development of social systems, revolutions, economic crises, changes in paradigms in science, and so on.

Despite the principled unpredictability of such unique events, at present there is a trend in attempts to predict them. In the run-up to 2017 and at its inception, a number of analytical papers by leading international and Russian expert agencies and organizations have been published on the possible scenarios for the emergence of "black swans" in international relations, the world economy and finance [6, 84].

Speaking about the category of uncertainty, it should be noted that it is the subject of system theory, along with the categories of *complexity* and *emergence*, which are traditionally considered as the main characteristic features of complex systems.

Despite the large number of works devoted to these categories, there are still no definitions of them that could be considered generally accepted, while the existing definitions are rather the nature of the descriptions.

In [144] it is noted that complexity is a common and natural property of artificial systems and takes place when it is impossible to trace a simple connection between the fact that there are elements of the system taken separately and the system as a whole; when the system has the properties of adaptation and the ability to achieve their own goals in different situations.

Complexity can be determined by the objective attributes of the system, such as the number of elements and their types, the connections between them or subjective assessments of the observer, according to his experience, knowledge, cognitive abilities and other factors.

One of the most common definitions of complexity [144] is the degree of difficulty in predicting the properties of the system as a whole with known properties of the constituent parts, which in turn is monotonically associated with the amounts of the elements of the system and the connections between them.

The definition of the complexity of the system through a measure of difficulty in understanding or describing (for example, calculations—then they say about algorithmic complexity, or about "Kolmogorov complexity"), or predictions of its behavior or properties are typical for many authors, in particular, [50, 127, 135].

Among the causes or sources of complexity, the structures of the elements of the system and their connections stand out [33, 51, 127, 131]; the possibilities of transformation of links [19], hierarchies [129, 130]; behavior and dynamic properties of the system [51, 127]; autonomy and relatively independent behavior of the elements of the system [47, 49]; wicked problems, unlimited context and unclear goals [33, 69, 81, 151].

In many works the complexity of the system is associated with the presence in its composition of people and their behavior (the possibility of active behavior, purposeful choice), political and socio-cultural factors [49, 127]. Much of the work is devoted to the complexity of biological systems, the evolution of species and natural selection [57, 58, 118].

In a number of works (see, for example, [152, 153]), Organized Complexity, which is inherent in systems of many strongly connected, ordered and diverse elements, demonstrating emergent, a priori unknown properties, is singled out and considered. Such systems are opposed to all the others, since they can not be described either by simple structures, by statistical laws, or by linear or equilibrium models. These include social, biological and similar systems that unite people, processes and technologies.

Emergence is defined (for example, in [27, p. 314]) as a property, according to which only when elements are combined into a single unit, properties that are not inherent in the elements are created. Emergent behavior should be considered as a consequence of the interaction and interrelationships between the elements of the system, rather than the behavior of individual elements. In [103], the concepts of emergence of two types are introduced: the "weak" emergence of planned and expected properties in the system, and "strong"—the manifestation of a system of unpredictable properties or behavior that are not observed until the system is created or modeled.

Summarizing the results of a brief analysis of the categories of complexity and emergency, one can conclude that practically all the factors inherent to them, firstly, are also characteristic of complex activity, and secondly, they are operationally manifested through uncertainty (including the properties and behavior of systems). Consequently, the category of uncertainty can be sufficient to adequately take into account the factors of complexity and emergence, therefore, in the development of CA models we will operate only with the category of uncertainty.

Since complex activity (as well as elementary) is inextricably linked with the subject of activity, the activity of the subject, his behavior, his purposeful choice is an important source of both true and measurable uncertainty.

Uncertainty of the CA can be generated by various sources associated with all the procedural components of the activity (Fig. 1.2). However, this influence is of a different nature, so we will group the sources of uncertainty and use it in the following discussion. To do this, let's consider consistently all the procedural components of activity (Fig. 1.2) and analyze the sources of uncertainty associated with each of them:

- *Uncertainty of the demand* is caused, on the one hand, by the uncertainty of the external demand for stakeholders from the side of the CA, on the other hand, by the person's perception of this need and the transformation of this "external" demand into his internal need.
- *Uncertainty of goals* and objectives depends, first, on the uncertainty of the need, secondly, on the goal-setting of the subject, and third, on the uncertainty of the external environment.
- *Uncertainty of conditions, requirements, norms and principles* is determined by factors of uncertainty of the external environment.
- *Uncertainty of technology and the subject* depends, first, on the environment (through conditions, requirements and norms), and secondly on the uncertainty (internal) means, methods and factors.
- *Uncertainty of actions and results* is secondary to the uncertainties of the subject, technology, subject and environment.
- *Uncertainty of evaluation, criteria, self-regulation*, in addition to external conditions of activity, depends entirely on the subject's active behavior and rational choice.

It makes sense to note the *activity* of the subject of CA, i.e. his ability in general to influence the external environment—the formation of external needs, standards, conditions, etc.

We group the sources of uncertainty by identifying the following primary groups:

1. Uncertainty (introduced by elements) of the external environment—external demand and external conditions, requirements and norms.
2. Uncertainty (introduced by elements) of technology and subject matter—means, methods and factors.
3. Uncertainty (active behavior and rational choice, including reflection) of the subject-awareness of the external need, goal-setting, implementation of actions, evaluation of the result, and finally (or, primarily) making a principled decision on whether to act as a CA subject.

In each of the groups there can be instances of both measurable and true uncertainty.

The problem of uncertainty is quite popular; the uncertainty is classified by different authors on the grounds:

- Description tool: probabilistic, interval or fuzzy;
- The source is a natural, game (resulting from the purposeful behavior of other actors), of a different kind.

We give illustrative examples of uncertainty:

1. Uncertainty of the external environment.

   True:

   - strategic uncertainty of demand, business, volume of demand for new banking services or for aircraft, etc.;
   - the emergence of extreme environmental conditions of significant fires and natural disasters affecting nuclear power plants (as was the case, for example, with the Fukushima nuclear power plant in 2011).

   Measurable:

   - the current unevenness of the mass demand, the business flow of customers to the bank branch during the day; flow of typical fire department calls;
   - current unevenness of environmental conditions, the intensity of traffic, making it difficult for the timely arrival of firefighters to the place of fire.

2. Uncertainty of technology and subject matter.

   True:

   - unique cases of equipment failure (as it was at the Three Mile Island NPP in 1979).

   Measurable:

   - refusals of standard equipment, defects in raw materials and materials.

3. Uncertainty of the subject:

   True:

   - creative solutions of the chief designers of the aircraft building company or "technologists" of the retail bank developing new financial technologies.

   Measurable:

   - typical traffic of the dismissal of employees of their own volition, illness of employees.

So, under the uncertainty of the CA, we will understand the possibility of a priori emergence of unpredictable events, the consequence of which is the inability to predict the result of the activity (its qualitative and quantitative properties, the moment of its receipt and the efforts expended). Uncertainty of CA will be divided into measurable and true; the external environment, technology and subject or subject; it is within the framework of this classification that we will consider it in the course of the subsequent presentation. The onset of such events generates new demand and/or new needs, and those in turn, new elements of activity.

2.8   Creation of Complex Activity Components

When implementing complex activity, the existence of some elements of activity is completed by satisfying the needs and does not entail any consequences. In other cases, in the course of the implementation or as a result of the completion of the elements of activity (achievement of the corresponding result), new activity, new elements are generated, when a new need is identified that is not yet satisfied with the activity carried out, or even the formation of a new need. After generation, CA elements are realized, completed and continue to exist in the form of obtained results and corresponding "knowledge", having the form of description of technologies and experience of obtaining the result—elements of the information model (Sect. 2.6).

Thus, one can speak of "*CA self-generation*" when one activity (some elements of CA) generates another CA (other elements of CA). The concept of "self-generation" is correlated with the ideas of H. Maturana and F. Varela, who introduced the concept of autopoiesis (Autopoiesis self-construction, self-reproduction, replication—that is, in some sense, self-organization, but not self-development) to cybernetics, see also [23, 24]. In [83], they defined a "self-replicating machine" (Autopoietic Machine) as a machine organized (and defined as a whole) in the form of a network of production processes (transformation and destruction) of components that: (i) through interaction and transformation continuously regenerate themselves and (ii) represent the machine as a specific unit in the space in which it exists, specifying the topological area of the space in which this network is implemented.

Generation of elements of activity occurs as a response to the onset of events of uncertainty, causing new needs/demand.

As a result of the onset of uncertainty events, external and/or internal conditions of activity change, which requires an appropriate reaction—a change in existing activity or the generation of new activity. Such an activity can be implemented with a known technology (when the conditions created by the onset of an uncertainty event are a repetition of what has happened before) or require the creation of a new technology (when the conditions are fundamentally new). The onset of events of measurable uncertainty generates new activity with a priori known technology, while true—often requires the creation of new technologies of activity.

Above sources of uncertainty were divided into three groups: the external environment, technology and subject or subject, they are also the sources of the generation of new elements of the CA.

Examples of generating activity are:

(a) in the course of the activity "the functioning of the branch of the retail bank during the day", the event of the external environment "the arrival of the client in the branch of the retail bank for the performance of utility payments (or the opening of a deposit)" generates new activity to provide this client with a specific service;

(b) in the course of performing the activity "designing or upgrading a certain aircraft model," the technological event "the need to develop a node and, possibly, the technology of manufacturing parts or assembling it" generates the corresponding activity. Or a new technical or technological solution found by the designer or technologist ("subject's uncertainty") generates activity to change the production or servicing of this model;

(c)   during the activity of "functioning of the fire department during the shift", the external event "fire detection" triggers the fire department (calculation) to extinguish it;

(d)   during the operation of "operation of nuclear power plants during the shift," the technological event "failure of specific equipment" generates activity for its repair or replacement.

Activity (its elements), as a rule, is generated as a response to *external* factors or events. These factors and events lead the subject to activity and are often a priori uncertain. The sources of these factors, as was shown in the previous section, can be external (in relation to the generated activity and its subject) environment, technology and subject or subject of other activity. Sequences or flows of such events create an *undefined demand*. Since activity is generated by external sources, users or *consumers* of the result of activity are also external persons (in relation to the CA and its subject). At the moment of generating activity, the individual actualizes external demand, turning it into his inner need, and at the same time becomes an elementary subject of CA.

In Table 2.3 shows illustrative examples of demand and users of the results. They quite clearly illustrate the thesis about the external character and demand, and the consumer of the result in relation to the activity.

In addition, consider an example where the source of demand, the consumer of the result of activity and the subject of activity is the same individual. This happens, in particular, when the individual has a fairly primitive basic food requirement and its subsequent satisfaction, or vice versa—creative initiative in the fields of entrepreneurship, science or art. Suppose that a specific individual at a certain moment had a need for food (need A). In order to satisfy this need, the individual takes out food from the refrigerator and cabinet, prepares specific dishes and eats them, satisfying the need of A.

Table 2.3 Examples of demand and users of result

Example	Demand	Users of result
Operation of a department in a retail bank during office hours	Flow of customers	Customers
Design or modernization of a concrete model of aircraft	Flow of events "design of a concrete unit"	Chief designer
Operation of a fire department during a shift	Flow of events "fire detection"	Individuals, municipal government, society
Operation of an NPP during a shift	Flow of events "failure of concrete equipment"	Management and owners of NPP

Obviously, the need for A can be satisfied in many ways (hire a cook who will prepare meals, buy ready-to-eat meals in a store, go to a restaurant/canteen/to friends, etc.)—activity of various kinds. The emergence of the individual's decision to prepare meals himself corresponds to the actualization of his (awareness of it) needs B to "prepare dishes." At that moment, the corresponding activity B appears, and the individual becomes the subject of activity B. Then he concretizes the need for B in the form of supposed dishes (goal-setting) and comprehends the technology of their preparation (structuring tasks and forming technology), possibly in parallel with the implementation of actions. Preparing dishes (performing actions), the individual, as the subject of activity B, evaluates the result and carries out self-regulation, seeking conformity of the dishes to the plan (requirements B and the goal of activity B). He does this within the framework of activity B, being its subject, until or until it reaches the goal (the dishes correspond to the plan) or does not exhaust the resources (time, products and energy).

After the individual (as the subject of activity B) realized the activity, got the result and achieved the goal—he finished the preparation of the planned dishes, he evaluates how much these dishes (the result of activity B) satisfy the need for A. Estimation of the result of activity B in relation to the demand A is carried out outside of activity B, and the individual is not a subject of activity B. The demand A is external demand in relation to activity B.

The ratio of the activity itself, the consumer of its results and demand is illustrated by the scheme in Fig. 2.6.

Schematic diagram Fig. 2.6 underscores the existence of two "managerial", reflexive circuits of feedbacks. One of the contours is realized by the subject of activity and includes the components of the activity and comparison of the result obtained with the planned goal. The second circuit consists of demand, activity in general, and evaluation of the result of the activity to meet its demand requirements, this contour is realized ("locked") by the consumer of the result of the activity. In a particular case, the subject of activity and the consumer of its results can coincide (as was the case with the satisfaction of the need for food).

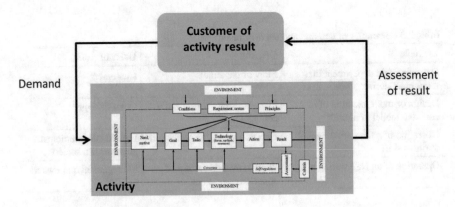

Fig. 2.6 Activity, its demand and results

Here there is a two-level comparison of the results of activity: first, with the need/goal of the CA and, secondly, with the requirements of demand. Multiple *verification* of the result corresponds to the system engineering concept [62] *validation* and verification of the system being created. The goal of the validation process is to check the implementation of the fixed requirements in the system created. While verification is aimed at obtaining reliable evidence that the system, when it is created, will meet the mission and needs of the stakeholders.

Once again, we emphasize that the generation of activity is closely related to the realization of uncertainty factors, both measurable and true. Also, what is very important, these factors and sources are external to the generated activity.

The foregoing allows us to formulate the assertion

External and uncertain nature of demand for complex activity results

Activity (its elements), as a rule, is generated as a response to external factors or events. These factors and events lead the subject to activity and are often a priori uncertain. Since activity is generated by external sources, the users of the result of activity are also external persons (in relation to the CA and its subject). At the moment of generating activity, the individual actualizes external demand, turning it into his inner need, and at the same time becomes an elementary subject of CA.

The system-wide features of the generation of CA, the demand and the consumer of the CA result are discussed in more detail in Chaps. 4 and 5, devoted to the corresponding models of CA.

2.9 Complex Activity Resources

Resources are one of the integral elements of both complex and elementary activity. They can be considered as a *means* of activity within a particular technology used or as a *condition* of activity [99] (the broadest interpretation: resources characterize the possibilities of carrying out activity). The problems of materiality of resource constraints, resource provision are the most important in the organization and realization of any activity, so consider the system-wide features of CA resources.

Let's consider several definitions: resource means, stock, opportunity, source of something. According to [61], "the resource is everything that is used in a targeted way, including this can be everything that is used for the targeted activity of a person or people, and the activity itself. A resource is a quantifiable possibility of performing any activity of a person or people; conditions that allow you to get the desired result with the help of certain transformations."

In various fields of knowledge, there is a large number of resource classifications; in the interests of our research, we will share resources based on their activity: the *animate* (individual, group, team, collective, organization) and *inanimate* resources (referring to the latter and animal resources), as well as on the basis of conservation and the form of application in the activity: *expendable* (materials, working time, energy, finances, etc.), *used non-accumulated* (personnel, tools, equipment, buildings, structures, etc.), *used accumulated* (or information—technology, knowledge and other similar objects that have a non-real form).

It is necessary to note the high importance of resources for activity in general and complex activity in particular:

- Resources provide the technology of activity and are its most important component;
- Pools of animate resources act as elements of a complex activity entity;
- Both animate and inanimate resources can form a CA item.

Resources in relation to activity have several significant system-wide properties:

First, most of the resources (other than accumulated or informational) represent, perhaps, the only real component, the component of the activity (see Fig. 1.2).

Secondly, because of the materiality, it is with them (as well as with the subject of STS) costs (costs for functioning in the process of activity, as well as costs for maintenance, maintenance and development) are associated.

Third, resources, as well as the subject-STS, are secondary to the activity. If resources are provided by a technology or CA subject, then their material forms are not needed, and their capabilities are qualitative (the ability to perform exactly the required operations with the object) and quantitative, volumetric (the ability to perform the required number of operations, or their volume). When an object of a CA is formed from resources, it is also not their current material form that is important, but their ability to transform under the influence of activity into the required result.

The concepts of life cycles are applicable to resources. In many cases, it is constructive to consider the LC of the resource pool, the resource instance's LC. For example, the concept of the life cycle is widely used in the definition and implementation of a strategy for managing productive assets, the concept of the *life cycle of an employee* is also used in the management of human resources, encompassing the processes of search, recruitment, development, as well as productivity motivation and dismissal of employees, taking into account evolution—the acquisition of experience and managing their competencies.

The life cycles of any of the considered resources, and their pools, and instances, it makes sense to consider as a way of organizing "auxiliary activity", the subject of which they are, therefore the resource center is also described by the steps introduced in Sect. 2.3.

Because of the need to maintain, maintain and develop resources, in many cases of complex activity, activity is created to *organize resources* and manage their life cycles.

The organization and management of the life cycles of resources in the system-wide sense boils down to:

- establishing and maintaining the composition and structure of the functionality of resources;
- providing resources for technology provision, the formation of the subject and the subject of the CA.

At the same time, quantitative and qualitative characteristics of resources should be provided at the levels specified by the activity for which the resources are intended.

The model of the resource pool is presented in more detail and is discussed in Sect. 5.4.

Organization and management of resource lifecycles is a special case of "auxiliary activity", the subject of which are resource pools, as opposed to "core activity", technology, the subject and subject matter of which provide resources. Resource pool refers to organized aggregates of resources—in interrelationships with "core activity" and with each other.

In the case of inanimate resources, the problem of organization and management of resource life cycles is considered and solved within the framework of standards, methods, procedures and software products for the *management of productive assets* (Enterprise Asset Management), for animate resources, resource organization refers to the area of *human resource management*, human capital, for information resources—knowledge management (see Sect. 2.6).

The problems of managing employee pools are closely related to the tendency of weakening organizational and structural ties and the transition to promising forms of organization of firms [32] in the form of pools of employees with equivalent competencies and the operational formation of working groups from them to carry out tasks. The organizational structure of the company begins to reflect to a lesser extent the managerial relations, and in the larger—to be the structure of the competencies of its personnel [156].

Animate resources—individuals have the ability to actively choose, and this specificity is taken into account when considering the appropriate models in subsequent chapters of the work.

The accumulated or information resources differ in two significant system-wide features. Firstly, being acquired, knowledge can be applied sequentially or in parallel without restrictions on time, space or the number of facts of application. Accordingly, pools of accumulated or information resources differ from pools of other resources in that their nature allows unlimited use or replication: they have no restrictions on the availability of the volume or the number of copies. Secondly, there exists a knowledge, inseparable from individuals who are carriers of this knowledge. Such information resources will be considered as specific competences of individuals, and the resources themselves associated with individuals with animate resources.

In illustrative examples, pools of animate resources are formed from client managers, accountants, loan officers, cashiers of a retail bank, engineers of certain qualifications and specializations of an aircraft construction company, workers of certain specialties of an NPP. Pools of used inanimate resources are, respectively, retail bank ATMs, the same type of equipment or buildings and structures of the aircraft construction company; pools of consumable resources—component parts and assemblies, warehouse stocks of fasteners of the aircraft building company, raw materials and materials, as well as energy in the case of other industries. Pools of accumulated or information resources are in all examples of technological information that exists in the form of documents, programs, database content and other forms.

The "core" activity that uses resources is external to the "auxiliary" and creates demand for it, in response to which resources and "auxiliary" activity, in turn, create restrictions on the implementation of the main. In many cases (primarily animate resources) the role of the subject of "auxiliary" activity is played by individuals or organizational systems that are also subjects of resource-using activity.

Due to the importance of resources, models of "support activity" for organizing resources and managing their life cycles are presented in the work; this section is devoted to Sect. 5.4.

The organization of resources is of great practical importance due to the materiality of the resource-related costs. To solve the problems of resource organization, a significant number of management methods, approaches, theoretical tools (in particular ERP, MRP and other concepts are briefly considered in Sect. 2.10), as well as software products (primarily SAP products [122] and Oracle [102],) for various types of human activity. Practical activity in resource organization make up a notable share of the global economy as a whole: the largest companies developing software ERP products, SAP (developing only ERP solutions) and Oracle (except for ERP systems developing other software products) in 2016 ranked in the largest in the world of companies 179 and 82 seats respectively [146].

2.10   Requirements for a Methodology of Complex Activity

The above motivated the authors to try to develop a methodology for complex activity: to create a common basis for managing activity, a unified system of CA models is necessary; factors of the weakening and disappearance of structures, globalization and integration of activity have increased the need to develop common approaches and models; intensive development of methods of artificial intelligence requires adequate formal means of supporting the application of these methods and especially the integration of human activity and the "activity" of the machine. Thus, MCA should serve as a basis for solving the above problems and, at the same time, reflect all the characteristics of the CA identified above.

Complex activity is a complex system, therefore, to determine the requirements for MFD as an integral system of CA models, we use system engineering methods [62], in particular *model-oriented* [42] and *architectural* [68] approaches.

The *architecture* of a complex system is defined [68] as a fundamental concept of how the basic properties and functions of the system are realized by the elements of the system in their interrelationships. Architectural descriptions reflect and express this fundamental concept and are widely used to create (design) systems, understand the essence of the system, its key properties, behavior and evolution.

The architectural approach [68] postulates that, due to the complexity of the system, it is necessary to use multiple models to adequately and completely describe it. Therefore, it is recommended to form an architectural description of the system in the form of a structured set of interrelated models that reflect different properties of the system. Also useful tools are *architectural templates*—agreements, principles and practices for the formation of architectural descriptions in a specific application area, primarily descriptions of enterprises and organizations. The development of architectural templates historically began with the now classic Zahman template [157], at the moment, dozens of templates have been developed and widely used, many of which have become international standards [95, 145]. Architectural templates are also created for branch [147, 148, 149] and highly specialized applications—see, for example, [9].

In the field of "system engineering" knowledge architecture is a definite analog of the methodology, therefore system-technical recommendations for the development and application of the architecture are useful in the development of MCA. We will consider below the methodology of CA in the style of an architectural template—as a set of models reflecting the main features of CA.

According to the model-oriented and architectural approaches, the CA model system should constitute an interconnected set of model-elements that represent the basic properties and aspects of the system, including its composition, structure, functions, behavior, requirements and parameters. At any time, any system is in a certain *state* and has some *structure*, which is the static aspect of the system. The processes reflect the *dynamics* of the system, describing how the system changes its state, and how its composition, structure, and functions change. *Complexity* of the system is reflected by means of decomposition and aggregation of structural elements and processes.

This mechanism allows you to focus on the details you are currently studying, leaving the rest in the form of a less detailed context.

We use as a preliminary basis the requirements for MFD system-wide characteristics and features of complex activity identified above (Sects. 2.1–2.9).

Let us briefly analyze the various fields of knowledge describing and investigating organizational and technical systems and similar objects related to CA with the aim of identifying opportunities to satisfy the preliminary basis of requirements, that is, in fact, we will justify the necessity of developing an MCA as a new theory. The results of the analysis are given in Table 2.4 (see also the reviews in [44, 48, 60]).

In accordance with the adopted approach to the construction of MCA (see the principle of generalization and abstraction above), based on the above-described features of the CA and the analysis of related areas of knowledge, we will form the final basis of the requirements that we will use later in the development of MCA. The basis of requirements is a preliminary basis; analysis of related areas of knowledge did not reveal new requirements, we can only supplement item "d" with the advisability of taking into account the value/usefulness of activity when analyzing the degree of achievement of goals.

Table 2.4 Brief analysis of different fields of knowledge related to MCA

Field of knowledge	Scope with respect to CA
Methodology [97]	Does not consider the subjects and subject matters of activity as well as its structure; therefore, does not provide a constructive hierarchical/network description and analysis of CA
Theory of systems [14, 106, 123]	Considers systems in general, does not offer models for theoretical studies of CA and its practical organization
Theory of organization [17]	Considers the general properties of the process and properties of organization; is one of the foundations of MKD, but does not offer models for describing CA
Cybernetics [7, 100]	Considers the general laws of control and communication in systems of various nature; is one of the grounds for considering managerial activity within MCA
Systems Engineering [62, 144]	Combines the approaches and methods for the creation, production and operation of complex systems
	Does not consider activity as a separate subject of research. Does not consider uncertainty, and hence does not offer tools to treat uncertainty
Control of organizational systems [101]	Is based on general methodology, but focuses on organizational systems, leaving activity in the background; does not consider multiple subjects and objects; does not describe the generation of activity
Theory of firms and enterprises (e.g., see a survey in [44])	Includes a significant number of theoretical concepts and methods: from A. Marshall's profit maximization [80] and the entrepreneurial theory of I. Schumpeter [125] to the theory of property rights [52] and information theory [4]
	Analyzes the system features of firms [143]. Substantiates the requirements of completeness, systemicity and dynamism of the unified theory of an enterprise, aimed at reflecting the integrity of its operation as a multifunctional and complex subject in the economic space and time
	Treats a firm as a set of assets; analyzes their complementarity [89], the processes of acquiring, processing and using information [4]
	However, considers firms and enterprises primarily as OTSs; pays almost no attention to activity, its features and characteristics

(continued)

Table 2.4 (continued)

Management and its branches (e.g., [45, 88] or the survey [60])	Considers various practical aspects of managing a company, enterprise, organization as an OTS:
	Strategic management (e.g., [3, 110]) as one of the management functions that applies to the long-term goals and actions of a company;
	Operations management, as the management of the process of purposefully transforming resources (materials, information or people) into target services or products of an enterprise. In particular, Sales and Operations Planning (S&OP) and its extensions—Integrated Business Planning, Advanced Sales and Operations Planning [25, 104, 105];
	Methods and tools for optimizing operations and quality (LEAN, TQM, 6 Sigma, 7S Framework McKinsey, etc.) Analysis methods for resource, process and organizational constraints and optimization: theory of constraints (TOC), critical path method. Reengineering methods of business processes [53];
	Management by Objectives (MBO);
	Management by Values, Value Chains [109];
	Organizational control as the management of organizational structures;
	In the presence of a very large number of works in this branch of knowledge, approaches and methods do not possess such an important property of any theory as generalization Goal-setting, activity in general, uncertainty, and life cycles are out of consideration Therefore, the approaches and methods proposed work well in the conditions of particular and specific (for the most part, repetitive) tasks. Their effectiveness and even applicability to completely new problems is a priori unknown
Theory of business processes [22, 64]	Is a method to specify and study business processes[a]
	Despite the development and rich variety of specifications of business processes, does not consider uncertainty; does not explicitly consider goal-setting, subject of activity (actor), and subject matter of activity

(continued)

Table 2.4 (continued)

Management of projects, project portfolios and project programs	Describes activity within a single project, project portfolio, or program. Considers uncertainty only in the framework of project risk management; does not describe the generation of new projects (activity). Does not offer any means of describing and analyzing "non-project" activity
Risk management [66]	Describes the risk aspect only and is not integrated into management; considers risks within an organization or product/asset/system. Has no connection with goals, and no comprehensive analysis together with opportunities, i.e., neglects the "positive" uncertainty of business.
Reliability theory	Considers the failures of technical systems associated with statistically measurable uncertainty
Operations research[b] [138, 150]	Deals with the development and application of methods for finding optimal solutions based on mathematical modeling. Has no explicit connection with goal-setting (goals are given) and activity. Can be used in MCA to optimize particular solutions

[a]Process is a set of interrelated or interacting activities that transforms inputs into outputs [64, p. 7]. Process is a stable, goal-oriented set of interrelated activities that transforms inputs into outputs that are of value to the user in accordance with a certain technology.Business process is a related set of functions in the course of which certain resources are consumed and a product is created (a tangible or intangible result of human labor: an object, service, scientific discovery, or idea) that is of value to the user.

[b]Operation is a set of actions or measures aimed at achieving a certain goal, i.e., set of goal-oriented actions

Applying these principles to the identified system-wide features of CA, we formulate the requirements for a unified system of CA models, which form the core of the methodology of complex activity:

(a) The MCA should include models of both elementary and complex activity, i.e. as having an internal structure, and not possessing. Hence, models of elements of activity and instruments for their integration are needed.

(b) The CA possesses a logical structure, in general a multi-level one. Since CA elements of different levels are themselves CAs, one can say that CA is fractal or self-similar. Structural models of CA must reflect its hierarchy, nesting and fractality.

(c) The technology of complex activity determines the cause-effect relationships between the elements of complex activity; hence, the MCA should contain cause-effect models of CA.

(d) Comprehensive activity is focused; therefore, the ICA should allow describing and analyzing the structure of the objectives of the CA, as well as the characteristics of the degree of achievement of the goals, the creation of value/utility as a result of the CA.

(e) Elements of CA exist in time: the needs for the results of CA arise, generate elements of CA, which are realized and then cease to exist. MCA should describe the life cycles of CA elements.

(f) The ICA should allow describing and analyzing the non-definiteness of the CA (measurable and true), realized in the form of the onset of a priori unpredictable events. The response to uncertainty (the onset of events) is the generation of a new activity (not before the event) with a known or requiring the creation of a new technology. MCA should describe the generation of new elements of CA.

(g) The MCA should describe the creation of new technology of activity, as a consequence of requirement f).

(h) The resources used, consumed and accumulated during the realization of the activity are an essential aspect affecting the technology, subject and subject matter of the CA. The MCA should contain models for the organization and use of resources.

(i) The MCA should include models of activity such as organization and management.

(j) Complex activity is implemented in the form of elements, each of which is conventionally related to process or project types, which must be taken into account in the respective models. MCA should combine project and process approaches within a single formalism.

(k) The CA methodology should represent all modern forms of organization of activity (see Sect. 2.3): (a) elementary forms and (b) complex operations; (c) projects and project programs; (d) life cycles.

(l) Multiple and complex links between CA elements and their subjects, the emergence of "meta-entities" of CAs that constitute "meta-organizational systems" (extended enterprises) are essential aspects of CAs that need to be taken into account within the framework of the MCA.

It is easy to see that the requirements (a) … (i) are the main ones—they define the composition of the MCA models and the key requirements for them, and the requirements (j) … (l) specify only additional requirements for MCA models.

Summary of This Chapter
The concepts of elementary and complex activity are defined; system-wide features of complex activity are formulated, i.e. features, which are characteristic for CA in any branches of human activity. *Elementary* it is proposed to name an activity, whose goals, technologies and the result do not have an internal structure (or the introduction of such a structure does not provide additional knowledge, effect).

Characteristics of elementary activity are:

(1) the technology does not change during the activity, the activity itself is clearly defined for the observer (researcher, subject of activity, consumer of its results) framework;
(2) the subject of the activity is unique and varies in its process according to the technology (which, in fact, constitutes the goal of the activity), but does not change its place and role in the context (requirements to the subject do not change);
(3) the entity carrying out the activity is unique and clearly defined and limited, and also not transformed in the process of carrying out activity.

Activity that is not elementary is defined as *complex*. Complex activity has a nontrivial internal structure, multiple and/or changing subject, technology, the role of the subject in its target context.

The general system characteristics and properties of CA are briefly considered:

• logical and causal structures of CA;
• the life cycle, as an essential factor in the realization of CA;
• the relationship of elements of CAs and their subjects;
• process and design types of CA elements;
• uncertainty of CA;
• generation and existence of CA elements in time.

The set of requirements to the methodology of complex activity that are used to create the system of models described below, which form the core of the MCA, is described. Appendix 4 shows the results of the correspondence of the constructed CA models to these requirements.

The *"ontology" of the basic categories* of MCA is shown in Fig. 2.7 in the form of a "substance-relationship" diagram.

Activity (its elements), demand, subject, technology, resources are characterized by their own life cycles, therefore the *problem of coordinated management of life cycles* (of demand, activity, its subject and subject matter, knowledge, technologies and organizations) is actual.

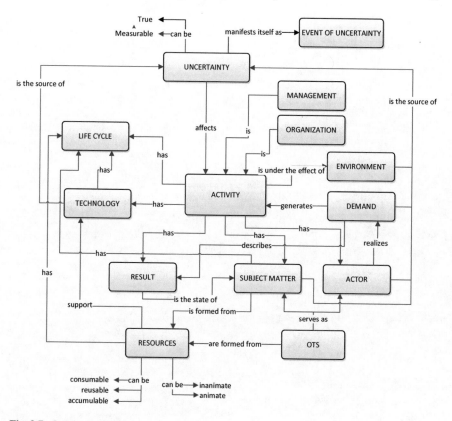

Fig. 2.7 Ontology of MCA basic categories

Chapter 3
Structural Models of Complex Activity

This chapter describes the models of the CA element, the logical and causal structures of the CA, the properties of the elements of activity are analyzed.

In the analysis of complex activity, the second chapter identifies its *main system-wide property*—the presence of a non-trivial internal structure, as well as structural fractality: the CA is decomposed into elements that in many cases are themselves CAs. Based on these properties, the model of the CA element and the model of *complexing* elements of CA structural models that allow recursively obtaining models of aggregated elements of higher-level CAs and CAs as a whole are considered below. Above, in the second chapter, it was stated that the distinction between logical and causal-temporal structures of CAs is essential; therefore, the models of recursive integration of CA elements are realized in the form of logical and causal structures.

This chapter includes four sections: the first one introduces the formal model of the element of complex activity (the structural element of activity, the so-called SEA), the second considers the properties of the CA element, the third and the fourth are devoted to structural models of CA—logical and causal, respectively.

3.1 Model of the Structural Element of Activity

Consider a model describing a CA element that meets the requirements a-1, formulated in Sect. 2.10.

As a basis, it is natural to take the well-known scheme [97] of the procedural components of activity—see Fig. 1.2 of Sect. 1.3. Part of the components shown in the figure refers to management, goal setting and constraints, being an "organizing superstructure" over another "implementing" part (see Fig. 3.1 and Chap. 7). An important factor external (in relation to the activity) environment is demand, which is perceived by the subject and actualized by him in the form of his own internal needs.

M. V. Belov and D. A. Novikov, *Methodology of Complex Activity*, Studies in Systems, Decision and Control 300, https://doi.org/10.1007/978-3-030-48610-5_3

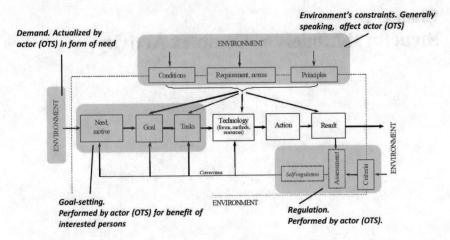

Fig. 3.1 The structure of procedural components of activity [97]

The procedural components of activity, shown in Fig. 3.1, reflect the "elementary *cycle of activity*", in which, according to technology, certain *actions* are taken leading to a certain *result* (which in general may not coincide with the *goal*, as with the desired, anticipated image of the result of the activity). Consecutive passage of this cycle—from the need to the result—will be conditionally called the *realization of the activity* or the realization of the element of activity (see below).

Concentrate on "implementing components", and "organizing" aggregate. "Implementing" components are technology, action and result.

In addition to these components, the CA element model should be supplemented with components representing the *subject* and *object* of the CA (which, strictly speaking, are not elements of the activity proper, but are intrinsic to it). The need for their inclusion is due to several factors.

First, in the introduction it was shown that the subject of this study is a complex pair-complex activity, as a primary subject, and STS playing the role of subject/subject as secondary, therefore, STS is also the subject of consideration.

Secondly, for the practical use of the CA methodology as a tool for building activity management systems and organizational and technical systems, the developed models should ensure the formation of requirements for STS, and for this, a CA subject must be included in the model.

Thirdly, the subject of activity is always active, and the properties of his active, moreover, reflexive, behavior require reflection and analysis. The subject of the activity (organizational and technical system) "is comparable with the researcher (with the management system) for perfection" [75], therefore it must be considered as part of CA models.

Fourthly, complex activity is a complex system, therefore, in its analysis, the most significant elements of the *external* (in relation to the activity) *environment* should also be considered: it is the subject realizing the activity and the subject of the activity, the change of which is the essence of the latter. The subject functions in conditions

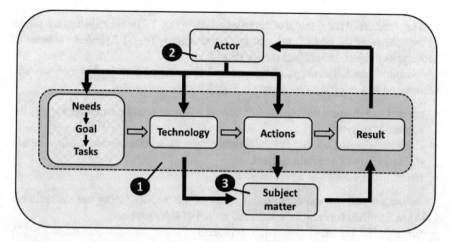

Fig. 3.2 The model of structural element of activity (SEA)

of environmental constraints and "translates" them into activity; also, the subject perceives the *demand* for expectations and interests of stakeholders, implements goal-setting and management of activity. Thus, the subject and the subject matter "encapsulate" all interaction of the external environment with activity as a system.

Fifthly, in most cases, a significant part of resources is spent on the change of the CA subject, therefore, in order to take into account the use of resources, the object of activity must inevitably be represented in the CA model.

Thus, we represent the model of the structural element of activity in the form of Fig. 3.2 [8, 10].

In terms of system engineering [67], activity (1) is a target system (System-of-Interest), objects (3) are systems expressing the operational environment (Systems Comprising Operational Environment), subject (2)—the system providing the target system (Enabling System). All elements (1-3) form a single system associated with the CA.

Each of the elements of the triad (1)–(3) is a system, so the triad (1)-(3) is a *system of systems.*

The arrows in Fig. 3.2 have the following semantics. The arrow from the subject to the target aggregate (need-goal-goal) reflects the fact that it is the subject that performs the goal-setting; from the subject to technology and actions—that the subject realizes the technology by a set of actions (acts accordingly technologies). The arrow from the result to the subject reflects the evaluation of the result, the self-regulation and the reflection of the subject.

The arrows from the technology and from actions to the object mean that the object changes as a result of the actions of technology, from the object to the result, that the result is the final state of the object, its evolution in the process of activity.

The structure of the element of complex activity (Fig. 3.2) is actually a composition of the scheme of the procedural components of activity (Fig. 1.2) and the scheme of activity, its subject and subject matter (Fig. 1.1).

We introduce the concept of **"structural element of activity"**, which we will understand as having the one shown in Fig. 3.2 structure object:

- created to achieve a certain goal/obtaining a certain result (transformation of the subject of activity),
- characterizing the activity (aimed at obtaining a result) in accordance with a certain technology over a certain subject,
- the subject of which is a certain STS.

Below we will use the structural element of the activity (using the abbreviation CSA) as a unified formalism, a standard model of CA elements.

Since the SEA abbreviation will be repeatedly used in the work with respect to CA elements in both the plural and the singular, and in various cases, we will add case endings and plural endings to this abbreviation, which, of course, is not traditional, but significantly improves unambiguous understanding of the text.

The formalism defined in this way (SEA) is a model of an element of complex activity. Due to the complexity of complex activity and the principle of complementarity, this model alone is not sufficient to describe all system-wide characteristics of CAs. Therefore, in the following chapters other types of models are presented. Structural models of CA provide modeling of non-trivial structures of complex activity. Process models reflect the realization of the stages of the LC of the CA elements, their dynamic aspects. The SEA formalism is used in all these models as a basic and universal element.

So, the elements of complex activity are described by the SEA formalism (see Fig. 3.2), followed by structural and process models.

We introduce a number of definitions, concretizing the category of the life cycle of a complex activity (the definition is given in Sect. 2.3) as applied to the CA element:

"Implementation of SEA" or *"Execution of SEA"* will be understood as a completed process, including awareness of demand and actualization of needs, goal-setting, structuring of goals and objectives, organization of the structure of CA (lower-level SEAs), technology formation, performance of actions (in accordance with technology) reception of result, reflection. "Implementation of the SEA" corresponds to the life cycle of the SEA (this is the time interval).

"Completion of the SEA" or "Obtaining the result of the SEA" will be understood as the completion of the actions in accordance with the SEA technology, obtaining the result and completing the process of reflection. "Completion of SEA" is not equivalent to achieving the goal of the SEA, because the result may not fully meet the goal (see above). "Completion of SEA" is an instant.

The concepts of "SEA" and "CA element" will be used interchangeably, if this does not lead to inaccuracies.

In the illustrative examples used by SEA, the following activity elements can be described (obviously, they are elements of different levels of hierarchy, complexity, etc., lists of examples do not in any way claim to be complete descriptions of some actual activity).

Examples for a retail bank:

- The activity of the bank as a whole during the financial year;
- The activity of the bank branch within a month;
- Issuance of an account statement by the employee to the client upon his request.

Examples for the aircraft building company:

- Implementation of the aviation program (for a specific model of the aircraft);
- Development of a certain technology for manufacturing parts;
- Assembling an aircraft (a specific product) in the final assembly shop;
- Performing the assembly of a particular assembly during assembly.

Examples for the fire department:

- Activity of the fire department during the reporting period;
- Conducting a regular fire-tactical occupation;
- Drawing up a fire extinguishing plan by the leader.

Examples for nuclear power plants:

- Station activity as a whole for a month;
- Activity of operators of the block shield during the shift;
- Execution of a planned detour by a station employee;
- Performing the installation operation during the repair.

Methodologically important for the correct construction of models is a particular case of a CA element that has a trivial structural and process representation, for its designation the term *elementary operation* is used. This case corresponds to elementary activity having a single goal that does not allow or does not require further detail, and in which all stages of the life cycle and/or all the procedural components, except the action, are degenerate. Elementary operation reflects the case when the fixation of demand by the subject means the beginning of the implementation of a single action, and the evaluation of the result as an independent stage is absent. A detailed examination of the stages of the life cycle of an elementary operation does not give additional knowledge, the entire LC can be considered a single stage "action and obtaining the result."

From the definition of an elementary operation it follows that its structural and process representations are trivial and consist of a single model element. In this sense, the elementary operation plays the role of a trivial element of activity in the methodology of complex activity.

Thus, any SEA is a composition of other elements of CA—SEAs and/or elementary operations, and any elementary operation is a "degenerate composition consisting of itself". And this applies both to the decomposition of the structural elements, and to the phases and stages of the life cycle of the CA.

Elementarity of the operation, obviously, depends on the point of view of the researcher. In accordance with the principle of abstraction and generalization, depending on the tasks being solved, in some cases it makes sense to abstract from the details and to consider rather complex CA elements as a single object, hence, in the model to describe them by an elementary operation, or vice versa, to detail intuitively simple elements for obtaining nontrivial conclusions. Description of the formalism of an elementary operation of obviously complex elements of complex activity is widely used in the following sections of this paper. All *system-wide models*, both structural and process, are constructed on the basis of this technique: abstraction from the specific features of CA elements by describing them by elementary operations and a detailed presentation of system-wide details with the help of the formalisms under consideration.

Let us illustrate the application of the definition of an elementary operation with the following examples:

– entering the details of the payment document into the bank's information system, or performing the posting on the accounts, or printing and signing the cash order executed by the bank's employee;
– fixing of a particular part on the place assigned to it during the assembly of the unit or the assembly of the aircraft by the workers of the aircraft building company;
– operation to dismantle a particular unit during the repair work of nuclear power plant equipment, carried out by the repair team.

The examples listed demonstrate CA elements that, depending on the purposes of the analysis, can either be represented as elementary operations, or detailed and represented by a multi-level hierarchy.

3.2   Characteristics of the Structural Element of Activity

The set of procedural components of activity in the form of Fig. 1.2 [97], used as the basis of the SEA Fig. 3.2, is a universal representation of elementary activity (not requiring further detail). We will show that the formalism of a single SEA allows us to describe elementary activity, having expressive properties not worse than other known formalisms.

First, in the theory of management of organizational systems [21, 86, 101], the scheme of procedural components is also used as a basis. 2, and its application is illustrated for the basic "subject-object structure of the control system". In [99] it was shown that the basic structure "is based on the activity scheme shown in Fig. 1.2, since both the subject and the control object carry out the corresponding activity, which can be described in the framework of the scheme of Fig. 1.2".

Secondly, elementary activity correspond to activity, for example, enterprises [144, section of Enterprise Systems Engineering] for the case of activity that do not require detailed elaboration.

It is easy to verify that the SEAs responds to each of the four enterprise definition options (Enterprise), as well as the ontology of enterprise-created value [144]. An important basic property of enterprises [144] is that their activity is aimed, firstly, at

Table 3.1 Different types of systems

		Control subject	
		Subject of activity	*Control device*
Subject matter	*Animate*	Active system, organizational system	Does not exist[a]
	Inanimate	Man-machine systems, ergatic systems	Automatic control systems

[a]It would seem that computer simulators can be placed here, as they put a person undergoing training in various situations of making decisions. But such simulators rather guide than control the trainee

creating a target value in the interests of external stakeholders (selling products and services to consumers), and secondly—on the transformation and development of the enterprise and its elements, the evolution of the object in the course of activity and the evolution of the activity itself in the course of its execution. Evolution and self-development are realized in this case through the evolution and self-development of the subject of activity.

Third, the SEA formalism allows us to describe the elementary component of a complex adaptive system (that is, it has not worse expressive properties than the formal agent of a single agent in the *multiagent model*—see also below), which according to [59] is defined as a set of interacting, autonomous, learning agents, decision-makers integrated into the environment interacting with them.

The formalism of SEA also uniformly describes active systems, man-machine systems and automatic control systems, as illustrated in Table 3.1.

In the latter case, it is not activity (as it is immanent only to man) that is described, but a control loop in which there is an automatic control device. This control device performs the "action" of the "technology", respectively, to obtain the "result" programmed during the creation of the device itself. In this case there is no independent goal setting (or it is also a priori programmed), which is performed by the subject in the STS.

An interesting feature of SEAs' formalism is that in it are separated by a "truly human" aspects of performance (they can only be carried out subject) and all the others who (sometimes theoretical and today almost) allow full automation—execution without human intervention, these aspects are included in the technology and implemented as actions (see Fig. 3.2).

So, it is possible to generalize the elements of the CA on the *subject matter of activity* (specific classifications can be infinitely many by the specifics of the activity itself). In accordance with Table 3.1, this classification may be made on the grounds "real/non-real" and "animate/inanimate":

(a) Material inanimate object of activity—products/systems/facilities, components, materials, raw materials, energy;
(b) Non-essential subject matter—knowledge, skills and information;
(c) Animate subject of activity people (individual, group, collective, team, organization);

It should be noted that combinations of the form a + b form complex *technical systems*; combinations of b + c—*organizational systems*; combinations a + b + c—*organizational and technical systems* or STS.

With reference to illustrative examples, it can be seen that in each of them there are different SEAs, the subjects of which are all variants and their combinations.

For a retail bank in the SEA "The opening of a new branch of the bank," the subject is « a + b + c » organizational and technical system as part of recruiting and training personnel, the system operation regulations (usually replicated a typical set of instructions and documents) and create complex software and hardware (software systems, computer technology, banking special equipment, storage, office equipment, communication system with the head office, etc.), room.

In the case of SEA "Development of technology for designing fuselage units using the new version of software product XX" of an aircraft building company, the subject is knowledge—an object of type "b".

For the SEA "Carrying out firefighting and technical exercises," the fire department is the subject of firefighters and commanders—an item of type "c".

The subject of the SEA "Replacement of the emergency water supply valve for steam generators" (for the nuclear power plant) is real—the valve and the feedwater system are of the type "a".

It is possible to classify SEAs by *forms of organization of CA* (see Fig. 2.5). In particular, in Sect. 2.3 above it was noted that complex activity can be organized in various forms; we correlate these forms with the formalism of SEA (Table 3.2).

Activity within each of the forms (F1-F4) form SEAs, we will designate as SEAs F1-F4. In practice, these forms form complex hierarchical structures (structures and hierarchies of SEAs are discussed in Sect. 3.3).

For example, the life cycle of such a complex capital facility as a petrochemical plant or a nuclear power unit includes the stages of design, construction, and the manufacture of basic equipment, further operation, modernization and decommissioning.

Each of the stages consists of one or several design programs, for example, the stage of the construction is implemented in the form of related programs of the direct construction of the facility on the production site and the production of the main equipment. Each of the programs includes several projects, and those in turn are

Table 3.2 Different forms of organization of CA

F1	elementary, unique, goals-operations-jobs-processes	Elementary cycle of activity
F2	complex operations: F2 = complex {F1; F2}	
F3	projects and project programs: F3 = complex {F1; F2; F3}	Completed cycles of productive activity [21]
F4	Life cycles: F4 = complex {F1; F2; F3}	Integral set of completed cycles

Table 3.3 Simplest decompositions of SEAs

subprojects, elementary and complex work with detailed detail right up to operations (for example, mounting a controlled valve, or drilling, or a pump).

Similarly, the life cycle of the release of a particular aircraft model includes design, production preparation, production and so on programs that are similarly detailed in projects, subprojects, operations and elementary operations.

In the hierarchies listed above, all the upstream and downstream elements (forms) are SEAs, so this structuring is fractal and self-similar.

Table 3.3 presents six simplest variants of decomposition of CA goals and the formation of the hierarchy of SEAs.

The superior SEAs are marked with the symbol "C", and the subordinate "A-1" … "A-K" … …."A-S". In each of the cells of the table one of the simplest decomposition options is presented.

The first column (cells I, III, IV) presents a sequential decomposition, when the subordinate goals should be implemented sequentially one after the other, in the second column—parallel.

In the first line (cells I and II) the subordinate SEAs "A-1" … "A-K" … "A-M" decompose the sequence of technological operations and actions of the superior SEA "C". Cell I shows a successive chain of activity when the result of the preceding "A-1" element in the chain corresponds to the initial state of the item of the subsequent "A-2" element.

The result of the finishing element "A-K" corresponds to the result of the parent element "C". Cell II represents a parallel decomposition when the result of the "C" element corresponds to the composition of all the results of "A-1" … "A-M".

In the second line (cells III and IV), cases are shown where the downstream SEAs are formed to create the technology of the superior SEA "C". The results of the elements "A-1" … "A-R" … "A-L" is a technological process—knowledge—or technological complex—material items of the superior element "C". Elements "A-1" ... "A-R" are realized sequentially, similarly to item I, and "A-1" ... "A-L" in parallel.

The third line (cells V and VI) represents the creation or modification of the elements of the subject (STS) of the superior SEA "C". Elements "A-1" ... "A-S" are implemented sequentially, similarly to items I and III, and "A-1" ... "A-Q" consistently.

Variants I–VI illustrate only the simplest cases of decomposition of activity; more detailed and generalized questions of the formation of the structure of activity are considered in the following sections of this chapter.

Another possible classification is the assignment of SEAs to project and process types, as described in Sect. 2.5, respectively.

Also, activity can be carried out as the *creation of products* (when the subject of the activity belongs to the entity of activity) or the provision of services (the subject of the activity belongs to the recipient of services). In the interest of this work, these two cases are not separated, because from the point of view of CA methodology, they are identical.

3.3   Logical Structure of Complex Activity

Complex activity is characterized by multiple goals, objects, technologies and subjects (for example, the production, sale and operation of new models of complex technical products—aircraft or cars—requires the creation of new technologies, and as a result of production complexes and organizational systems), and such an interconnected set of CA components must be considered in a single logic.

The multiplicity and heterogeneity of components generates a complex structural organization of CAs, hierarchies of organizational forms and/or elements of CAs SEAs arise: any program for the creation and production of a new technical product—aircraft, car, mobile phone—includes a number of projects, each of which is decomposed into subprojects, operations, operations. SEAs are in a hierarchical relationship of responsibility/subordination, when the subject of the higher activity element is responsible for the performance of all the downstream elements—for example, the program manager of the petrochemical complex construction is responsible for all design and construction and installation works performed by various organizations.

First of all, it is necessary to choose the basis for distinguishing the structure of activity. Indeed, the *structure* is a set of stable links between the elements of the system. There are heterogeneous links between SEAs, and the choice of one or another type of relationship generates a corresponding structure. Complex activity and SEAs have different procedural components (see Figs. 1.2 and 3.2), each of which in many cases forms hierarchies and is potentially suitable to serve as a basis for identifying the appropriate structure of SEAs.

Any activity "starts" with the *need* that generates the *goal*, and ends with the *result* (both for the elementary operation, and for the CA of any level of complexity). The need, goal and result are the main "target" components of the activity, therefore, we will define the structure of CAs precisely by their structure, i.e. *structure of goals* (since the result obtained does not always coincide with the goal, and the needs are not always specific, and not they, but the goal defines tasks, technology and actions). Technology is secondary to tasks and objectives, so the technological structure (being, nevertheless, more detailed) corresponds to the structure of goals. Tasks are the decomposition of goals, their specification in specific conditions of activity. Actions, being aimed at solving problems within the framework of the technologies used, also follow the structure of their goals in their structure. The final result is a consequence of actions and technology, so its structure reflects the structure of the goals (taking into account the uncertainty inherent in the result).

Thus, the structures of other components of the CA "repeat" the structure of the goals, so we will use it as the basis for the logical structure of the activity—the composition of the SEAs and the "target" links between them, i.e. structure of SEA, which we will conventionally call the *logical structure of CA*.

We will show that attempts to use other grounds for isolating the structure of a CA on the structure of an object or subject violate the unity of its consideration. This happens, for example, in the case where a technology is created within the CA for performing any elements of the CA. This example is very common in practice, it occurs in case of any modernization of existing or creation of a new activity (or the corresponding STS). For our illustrative examples this, for example:

- the introduction of a new banking product (for example, Internet banking in general or an Internet service);
- development of a new production technology for the units during the development of a new aircraft model (for example, Boeing, when creating the Dreamliner model, together with Mitsubishi Heavy Industries, developed and introduced a technology for mass production of wing planes made of composite materials);
- introduction of new fire extinguishing means (for example, special unmanned aerial vehicles);
- methods of maintaining production assets (for example, tools and methods for monitoring metal equipment of the primary circuit of nuclear power plants).

In each of the examples there are elements of activity related to the development of new technologies, and then with the training of personnel to these technologies. In these cases, the structures of CA subjects and CA objects are violated, since the subject element of one CA element becomes an object for another CA element. Therefore, if to form the CA structure as a whole, then the "complex" subject will include (partially) the "complex" subject and vice versa, which is not productive and can often lead to confusion.

The logical structure of the CA is defined as a finite acyclic graph (Fig. 3.3), reflecting the fact that each SEAs (and CA as a whole, as a special case) is decomposed into a finite number of J subordinate SEAs and L elementary operations. In this case,

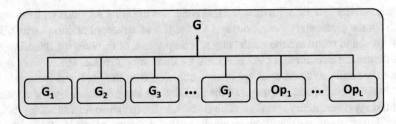

Fig. 3.3 The logical structure of SEA

the subject of all elementary operations, into which the SEA is decomposed, is the subject of the SEA.

The formation of hierarchical SEA structures, the decomposition of some SEAs into a set of lower-level SEAs and elementary operations, reflects the system-wide properties of CA; specific features are concentrated in elementary operations, namely, their content distinguishes one particular CA from another, having the same structure.

Since the logical structure reflects the goal decomposition, the graph has a single root vertex (in each decomposition operation, the decomposable vertex is unique). This means that to achieve the goal of G_N it is necessary to ensure the achievement of subgoals G_1, G_2, ..., G_J (the arrows are directed from subgoals or operations to the target) for which the corresponding SEAs are defined, as well as a number of sub-goals O_1 ... O_L obtained by performing elementary operations.

Not in all cases (not for all SEAs) the achievement of all subgoals is mandatory to achieve the ultimate goal of G_N. Uncertainty and generation of new elements of the CA explain this feature, which is discussed in detail in Chap. 4.

To ensure the correctness of the formalism, Fig. 3.4 presents the logical structure of an elementary operation, which has a trivial form. The arc-loop is conditional, illustrating the possibility of further "decomposition" of O_L only into itself.

The logical structure also reflects the relationship of the source of demand (it is also a consumer of the result of the activity, as noted above) and the subject of activity, considered in Sects. 2.2, 2.8 (Fig. 3.3 actually represents a fragment of the hierarchy): the superior SEAs and its subject demand for lower SEAs—set the requirements for the results of the subordinate and are the consumer of their results. In response, lower SEAs, having certain capabilities to meet demand, impose

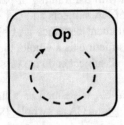

Fig. 3.4 The logical structure of an elementary operation

Table 3.4 Relations between subject matters and subjects of SEAs

		Upper SEAs		
		Material product	Knowledge	OTS
Lower SEAs	Material product	Creation of product's components	Creation of products (equipment) for technology	NO (indirectly through knowledge and technologies)
	Knowledge	Creation of industrial process for product	Creation of components of industrial process	Creation of operation technology of OTS
	OTS	NO (indirectly through knowledge and technologies)	Training of employees for executing industrial process	Transformation of OTS elements

restrictions on the implementation of higher. Thus, the connections of the logical structure also reflect the bilateral relations "demand—opportunities (satisfy it)" (or "*needs—opportunities*" in terms of [120]).

In general, the target structure of complex activity is not unambiguous. The formation of a logical structure and the optimization tasks to be solved are discussed in Sect. 7.2.

Relationships of higher and lower SEAs in terms of the nature of objects and subjects of their activity can be illustrated 8.

The columns correspond to the higher (senior) SEAs, and the rows to the lower (subordinate), the cells reflect the content of the relationship between the senior and subordinated SEAs. Subjects downstream in the hierarchy of SEA can be any of the elements of the superior STS, that is, lower SEAs can be organized to create/execute/transform the subject/result, technology or subject (STS).

Table 3.4 demonstrates that knowledge—the description of technology is the link between animate (personnel/STS) and inanimate (product/technological complex) elements and objects of activity.

Fractal properties of CA elements generate multi-level hierarchies of SEA (Fig. 3.5).

Logical structure of the CA in addition to the structure of goals reflects the "management" *hierarchy of the subordination* of SEAs and the responsibility of the subjects for the results—for achieving the goals. The subjects of the higher SEAs, together with the responsibility for achieving their "own" goals, are also responsible for achieving the lower goals, and, as a consequence, for actions within the framework of the lower SEAs. If any goal is the goal of the lower level in relation to only one goal, then the corresponding superior (directly) SEA is responsible for achieving the goal of the corresponding SEA. For example, in Fig. 3.5 are pairs of goals w and G_J, v and G_J.

Obviously, the structure of goals is not always tree-like, and there may be subgoals that are needed simultaneously for several higher-level objectives (in Fig. 3.5, for

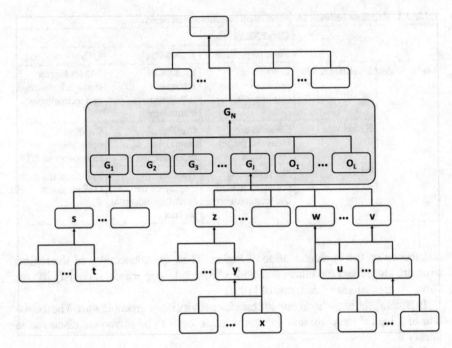

Fig. 3.5 An example of logical (goals) structure of CA (the structure of SEAs decomposition)

example, it is the vertex u relative to the vertices v and w), as well as the goal connections of "different levels" (x is a sub-goal for y and w, and t is a sub-goal for s).

In this case, the *principle of one-man management* determines that responsibility for achieving such goals (similar to u) is borne by the subjects of the SEAs, in whose area of responsibility are all the goals (v and w) whose sub-goals are the original goal (u). For the goal u, an SEAs (in whose area of responsibility both v and w lie) is the G_J SEA, as well as all SEAs that are higher in relation to G_J. It is natural to assume that the responsibility for the goal u should be carried by those SEAs, which are "closest" to u in the hierarchy, i.e. SEA G_J. The goal structure of CA can be described and analyzed using the apparatus of graph theory.

Foundations of logical structure of complex activity

The only constructive basis for structuring complex activity-the selection of the hierarchy of its elements-is the structure/hierarchy of the objectives of the activity.

The logical structure also reflects the "managerial" hierarchy of subordination and responsibility of the subjects of the elements of the CA for the results—for

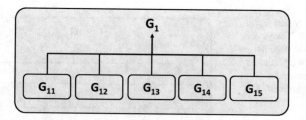

Fig. 3.6 The logical structure of the "Pilot introduction of a new banking Internet service" CA

achieving the goals. The subjects of higher SEAs, together with the responsibility for achieving their "own" goals, are also responsible for achieving lower goals, and, as a consequence, for the activity of lower-level SEAs.

Consider the logical structures of some illustrative examples.

The logical structure of the "Pilot introduction of a new banking Internet service (Goal-G_1)" CA is presented in Fig. 3.6.

Goal-G_{11}—development of a financial model for a new service;

Goal-G_{12}—development of technology (business processes and regulations) for the performance of the service;

Goal-G_{13}—development of a software product supporting the service;

Goal-G_{14}—conducting a marketing company to promote the service;

Goal-G_{15}—training of employees of the new technology.

Full-scale implementation of the service includes the following goals (see Fig. 3.7):

Goal-G_1—pilot implementation;

Goal-G_2—verification of design decisions during experimental operation;

Goal-G_3—finalization of technology and software product;

Goal-G_4—employee training.

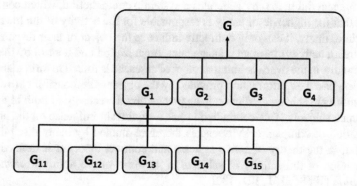

Fig. 3.7 The logical structure of the "Full-scale implementation of the new banking Internet service" CA

Fig. 3.8 The logical structure of the "Carrying out works on the current maintenance of a particular unit of equipment of a turbine hall of an NPP" CA

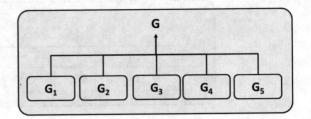

Element of activity "Carrying out works on the current maintenance of a particular unit of equipment of a turbine hall of an NPP" may include the following goals (see Fig. 3.8):

Goal-G_1 withdrawal of the unit from the operating mode;

Goal-G_2 dismantling of the unit;

Goal-G_3 diagnostics of assemblies and parts;

Goal-G_4 replacement parts;

Goal-G_5 assembly and testing under load.

Similar logical structures can be constructed for other illustrative examples.

It makes sense to clarify that the *management hierarchy* (the relation of responsibility) in relation to complex activity and its logical structure means the need for the subject to organize activity, monitor its implementation, to solve all the problems that arise while it is in the zone of its responsibility. On the other hand, subordinate subjects (subjects of lower-level SEAs) are obliged to follow the instructions of higher-level ones and provide higher-level complete and timely information, including information on emerging problems that require decision-making at the level of higher-level entities, and *escalate problems* to a higher hierarchy level. Thus, the management hierarchy defines the relationship of responsibility between the subjects of SEA, the relationship of nesting, the ownership of areas of responsibility, areas of escalating problems.

The managerial hierarchy generates a systemic contradiction, which consists in the fact that the higher-level entity is responsible for the activity of the lower-level ones, while, firstly, it can only indirectly influence the object of their responsibility (downstream activity) through management, organization and control by the latter, and, secondly, in the overwhelming number of cases, he is forced to form managerial influences, based on information provided by the object of responsibility (by subordinate entities). This contradiction is aggravated by the activity of both higher and subordinate subjects. This contradiction has a significant influence on the practical organization of complex activity and is therefore studied by many disciplines, in particular, in the theory of active systems and contract theory it is defined as the manifestation of the activity of subjects in the form of the ability to *manipulate information* [18, 86, 101, 121, 132].

The proposed logical structure of the CA is a distant analogue of the *organizational structure* of the firm, the project and other STSs. Therefore, for the description and optimization of the structure of the decomposition of STS, known results of optimization of hierarchical structures can be used [21, 92]. On the one hand, the logical structure of the CA, like the organizational structure, reflects the relationships and relationships of responsibility and subordination, on the other, unlike the organizational one, it does not record the relationship of belonging and the occurrence of some orghedinits in others. In addition, what is very important, the logical structure, unlike the organizational one, reflects the uncertainty of the CA and is "not strict" (see above in this section the observation that all subgoals are not necessary in the general case).

3.4 Cause-and-Effect Structure of Complex Activity

Complex activity has the structure described in Sect. 3.2: the goal of each SEAs is decomposed into sub-goals of the subordinate level, between which there are causal relationships. CA technology determines the cause-effect relationship between the objectives of CA-SEA and elementary operations: the same design, installation, repair and other works must be performed in a certain sequence to obtain the desired result. On the other side of complex activity, there is a significant uncertainty, which also affects the sequence of CA elements.

We introduce the cause-and-effect model of the SEA for describing the cause-effect relationships between the lower-level elements of the CA, taking into account the possibility of occurrence of uncertainty events. In Fig. 3.9 shows an example of the cause-and-effect model of the SEA, presented in notation corresponding to the basic part of the BPMN-notation [22]. The description of the causal model can be performed using various notations; BPMN-notation was taken as a basis for several reasons:

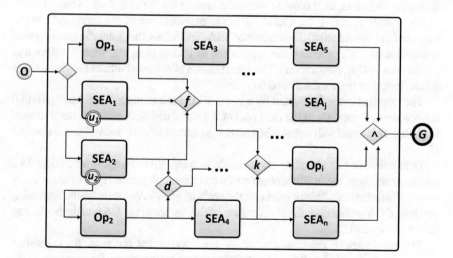

Fig. 3.9 Cause-and-effect models of CA

First, it uses concepts and basic elements (which will be briefly discussed in this section below), necessary for describing SEAs.

Secondly, it has developed expressive possibilities.

Thirdly, it relies on the system of generally accepted international standards ISO 9000 [64] and, at the same time, is itself supported as an international standard by the authoritative OMG group (see http://www.omg.org).

We describe the main elements and rules of BPMN notation, applying them to the model shown in Fig. 3.9.

Rectangles with rounded corners denote subordinate elements of activity—SEAs (SEA$_1$, SEA$_2$, ..., SEA$_J$) and elementary operations (Op$_1$, Op$_2$, ..., Op$_L$); circles—events of uncertainty occurring during the execution of elements of activity and affecting the course of its implementation; rhombuses (for example, d, f and k) are control points (gateways) reflecting branching (parallelization) and merging the execution of the elements of activity. Elements of notation are linked by lines with arrows, which reflect the cause-effect relationships and the order of performance of the elements of activity.

Just as in BPMN-notation, we use the theoretical concept of "token" (token) [22] (in petri nets their analogs are markers). In accordance with this concept, the execution of actions is reflected by the emergence and transfer of abstract objects, "tokens", from one element of the causal model to another, following the arrow-links.

The beginning of the entire SEA as a whole (Fig. 3.9) is modeled by the appearance of a single token in the model element that reflects the initial event (indicated by 0), the execution of the elements following the initial event (in this example, parallel execution of Op$_1$ and SEA$_1$) will correspond to the disappearance of the token in the "event 0 "and the generation of tokens in the elements Op$_1$ and SEA$_1$. Upon termination of Op$_1$ and SEA$_1$, the tokens disappear in them and are generated in the following elements, as if moving along the lines of the arrows. And so on.

The events u_1 and u_2 correspond to the implementation of the uncertainty in the course of the execution of the elements of SEA$_1$ and SEA$_2$, the generation and parallel realization of CA elements—the appearance of tokens in u_1 and then in SEA$_2$, and u_2 and then in Op$_2$, respectively. The specification of the cause-effect model includes a description of the events u_1 and u_2.

The terminal event (indicated by a circle with a thick line and the inscription G) reflects the achievement of the final goal of the simulated SEA and the completion of the SEA. This event will occur when all the generated tokens "move" to the element G.

Generation of CA elements corresponds to qualitative changes in activity. As a source of changes, the fulfillment of certain conditions, the occurrence of uncertainty events, "bifurcation", "black swans on the scale of SEA". As a result—the branching process of CA realization: the continuation of the "generating" CA and the beginning of the "generated."

The semantics of the causal structure presuppose that the need for consistent achievement of goals is reflected by their consistent arrangement, for example, of the goal of SEA$_3$ and SEA$_5$. The conditions for achieving goals that are a combination of several parallel goals are given by the corresponding rules (conjunctive-disjunctive forms). For example, in order to achieve goal f, it may be necessary to achieve a pair of goals for SEA_1 and Op_1 or for the goal of $SEA_2((SEA_1 \wedge OP_1) \vee SEA_2)$. In this case, the sequence of achievement of SEA_1 and Op_1 is not regulated, since in this case they are represented in the diagram as "parallel" objects. After any of the intermediate goals are fulfilled, the fulfillment of the goals—SEAs and elementary operations following it—begins. For example, after fulfilling the goal f, the SEA_j will begin, and after SEA_4 the SEA_J will begin.

Complexity of activity implies the possibility of parallel execution of many of its elements; therefore within the framework of the SEA, several downstream SEAs and elementary operations can be executed in parallel. This is reflected both by the presence of several outgoing arrow-links from the elements, and by the possibility of the occurrence of undefined events (in Fig. 3.9 u_1, u_2), which can be initiated during the execution of the SEA through the mechanism of the reaction to uncertainty, which will be described in detail in Chap. 5, devoted to process models.

The causal (cause-and-effect) structure reflects the technological links of the complex activity, being actually a system-wide CA technology, while the elementary operations (their content) and the cause-effect relationships themselves represent a specific part of the CA.

Causal model of complex activity

The causal model of complex activity reflects the specific factors of its technology.

The rationale for the validity of this statement is almost trivial: if specific links between elements (hence, logical and causal structures) are not traced, then such an activity makes sense to be considered as a set of independent elements of activity, which greatly simplifies the task.

We briefly formulate the basic rules connecting the logical (vertices—SEAs, arcs—private-general relations, logical conditions of attainability of the goal depending on the achievement of subgoals) and cause-and-effect (vertices—SEAs and elementary operations, arcs—precedence-following relations) model CA.

First, all the vertices of the logical structure, which are goals or subgoals, must be reflected in the causal model for setting the cause-and-effect logic of their achievement.

Secondly, the root vertex of the logical structure coincides with the terminal event of the cause-effect model.

Thirdly, the arcs of the logical (relations of the decomposition of goals and subordination) and the cause-and-effect (causality and following relations) structures have in general a configuration that is independent of each other.

The formalism of the causal structure must be supplemented by the trivial structure of the elementary operation (see Fig. 3.10) in the same way as was done in Sect. 3.3.

The fractality of the SEA is also manifested in the formalism of the cause-effect model: the aggregation of SEAs is illustrated by Fig. 3.11, where each of the SEAs

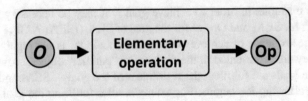

Fig. 3.10 The causal structure of elementary operation

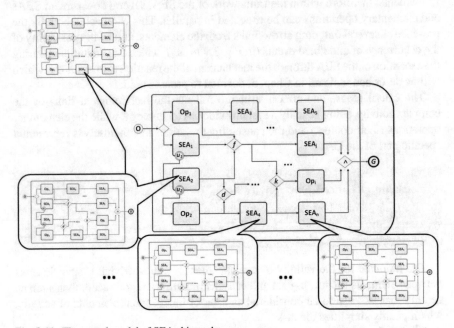

Fig. 3.11 The causal model of SEAs hierarchy

is represented by a more detailed cause-effect model (BPMN diagram) at the lower level of the hierarchy.

Let's consider the cause-effect structures of the individual SEAs from illustrative examples.

Pilot implementation of a new banking Internet service can be represented by the following causal structure (see Fig. 3.12):

G_{11}—development of a financial model for a new service;

G_{12}—development of technology (business processes and regulations) for the performance of the service;

G_{13}—development of a software product supporting the service;

G_{14}—conducting a marketing company to promote the service;

G_{15}—training of employees of the new technology.

Full-scale implementation of the service may include the following SEAs (see Fig. 3.13):

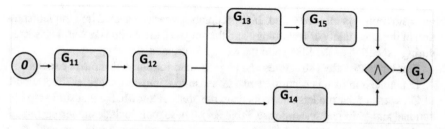

Fig. 3.12 The causal model of the pilot implementation of a new banking Internet service

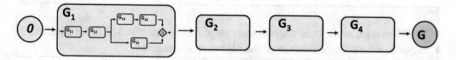

Fig. 3.13 The causal model of the full-scale implementation of a new banking Internet service

G_1—pilot implementation;
G_2—verification of design solutions during the trial operation;
G_3—finalization of the technology and software product;
G_4—employee training.

Carrying out works on the current content of a particular unit of the turbine hall equipment may include the following elements of the causal structure (Fig. 3.14):

G_1—output of the unit from the operating mode;
G_2—dismantling of the unit;
G_3—diagnostics of assemblies and parts;
G_4—replacement parts;
G_5—assembly and testing under load.

The diagram Fig. 3.13 reflects the sequential order of the implementation of actions by SEA 1, 2, 3 and 5. It is assumed that the implementation of element 4 (replacement of units and parts) may not always be carried out: with an acceptable

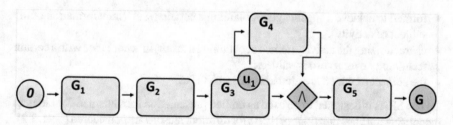

Fig. 3.14 The causal model of carrying out works on the current content of a particular unit of the turbine hall

wear, no replacement is required. If unacceptable wear is detected (the implementation of the uncertainty of technology and the subject), the initial vertex u_1 is initiated during execution of the SEA_3 and the SEA_4 is executed.

Similar cause-effect structures can be presented for other illustrative examples.

In conclusion of this section, it makes sense to compare the cause-effect model of CA with similar models and tools from the areas of knowledge that study organizational and technical systems (see Table 2.4). These tools include numerous formal models of the *theory of processes*, production and business processes, and, as a particular, but very important case, *network graphics* (and their various variations).

The causal model of CA is similar to the mentioned tools in the sense that, like them:

- is a structural model and reflects the causal relationship between the elements,
- supposes the complex nature of the elements themselves,
- uses the models of graph theory.

However, it differs semantically from the network graphs and the tools of the theory of processes:

- First, the SEA formalism is more "rich" (it represents the goals, technologies, actions, subject and subject matter) in comparison with the elements of the compared instruments (which describe actions—work, processes).
- Secondly, the natural property of the cause-effect model is its variability in the course of implementing the CA, described by it. In fact, it "changes itself", which is caused by the basic properties of CA—the uncertainty and generation of elements, their evolution over time (see Chap. 4).

In fact, the causal structure is the result of the deployment of a logical structure in time, taking into account technological links and resource constraints.

Summary of This Chapter
Models of the CA element (SEA), logical and causal structures of CAs are introduced; the properties of the activity element are analyzed.

The SEAs is defined as having the one shown in Fig. 3.2 structure object:

- formed to achieve a specific goal/obtaining a certain result (transformation of the subject of activity),
- characterizing the activity (aimed at obtaining a result) in accordance with a certain technology over a certain subject,
- the subject of which are elements of some OTC.

The SEAs is designed to be used as unified presentation formalism, a CA element model. Several classifications of SEAs for different reasons are considered (Sect. 3.2).

The model of the logical structure of complex activity is built on the basis of the objectives of the CA. It is shown that the most productive is the use of CA targets (and not other components thereof) as the basis for isolating the logical structure of CA.

A model of the cause-effect structure of the CA is proposed.

The logical structure of the CA reflects, among others, a hierarchy of management and responsibility links, while a cause-and-effect hierarchy of technological links.

Models of logical and causal structures of CA provide recursive integration of its elements, responding to fractal properties of complex activity (Figs. 3.5 and 3.11).

At the end of Sect. 2.10, a list of requirements was formulated, which should be answered by the developed methodology of complex activity and the system of models. Models of SEA, logical and causal structures are the main ones and directly ensure the fulfillment of requirements (a)… (d) and also support the fulfillment of all other requirements (e)… (l).

After considering the structural features of the CA, it is necessary to proceed to an analysis of the group of requirements for the methodology of CA associated with the realization of complex activity and its existence in time. This is the uncertainty and generation of the elements of activity—requirements (e)… (g), which are discussed in the next chapter.

Chapter 4
Uncertainty and Creation of Complex Activity Components

One of the most important system-wide features of complex activity is the uncertainty and generation of new elements of activity; in many cases they are directly related, so this chapter is devoted to their analysis and discussion.

The material of the fourth chapter includes five sections. In the first of them, the classification of SEA on the basis of the emergence of new elements of CA is introduced, the next three consistently analyze all the selected classes of SEA taking into account the uncertainty and activity of subjects (as one of the significant sources of uncertainty). In the fifth section, a generalizing model of the manifestation of uncertainty in the course of implementing the SEA is proposed.

4.1 Classifying Types of Activity and Its Structural Elements

In the realization of complex activity, some elements of activity are completed by satisfying the demand/needs and do not lead to any a priori unknown effects in terms of activity. In other cases, during the implementation of some SEAs, new SEAs and hierarchies of SEAs are generated under them: within the framework of execution of some SEA, there is a demand, then within the other SEA, the new requirement is being updated, which is not yet satisfied with the activity or even the formation of a new need. The emergence of demand can be caused by one of two factors (according to the results of Sects. 2.7 and 2.8):

- execution of deterministic decomposition of an already existing SEA according to the known technology;
- the occurrence of uncertain events causing the generation of elements of activity (analysis and examples of generation of elements of activity were given above in Sect. 2.8).

M. V. Belov and D. A. Novikov, *Methodology of Complex Activity*, Studies in Systems, Decision and Control 300, https://doi.org/10.1007/978-3-030-48610-5_4

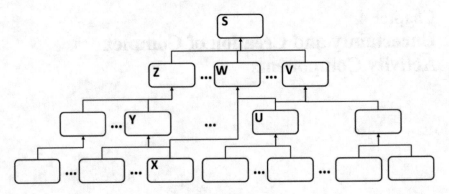

Fig. 4.1 The generation of the CA elements (logical structure)

The identified need leads to a new goal-setting—the organization of a new activity—the formation of a new SEA and the hierarchy of subordinates in the logical structure of SEAs: "new activity" can be directed either at satisfying the need, or in identifying the expediency (for the subject of activity) of meeting the need (for example, checking the profitability of an order). As the elements of activity are completed, the result is achieved, the SEA ceases to exist.

The order of the generation of elements of activity can be described as "generation from the top down": from the higher (in the logical structure) SEA to the lower ones. It makes sense to note that the postulation of the composition and logical sequence of the procedural components of activity (Fig. 1.2) leads to the fact that the order of "top-down generation" is the only possible one. We will show this, using the principle of proof "from the opposite", for which we consider the generation of the elements of activity that form the logical structure presented in Fig. 4.1.

The logical structure, as defined in Sect. 3.3, reflects the structure of the goals and structure of the SEAs. The SEAs, denoted by the letter S, is the root of this hierarchy, and the goal that must be achieved as a result of its execution is the main one. The demand that generates this structure describes the requirements for the results of the SEA S.

Let, for some reason, the first one is generated by an SEA designated by Y. But for this goal its own goal and, as a consequence, the requirements to its result must be defined. The goal of Y is one of the sub-goals of goal Z. In the absence of the superior SEA of Z objective sources, the goal of SEA Y and the requirements for its results are not. This means that the demand for the results of the SEA Y will not have constructive grounds; therefore, such a generation is not justified and expedient.

Similarly, we show that the "self-organization" of complex activity also necessarily has a basis for generating "top-down".

Let a priori exist SEAs Z, W, Y, and at some point each of their subjects somehow decided that it would be advisable to combine these elements of activity within the framework of some integer, integrating element, not yet of the existing SEA S. Since complex activity and its elements are systems, to create a new system (element S), it is necessary to "someone" who will comprehend it (the future SEA S) as a whole, and certainly greater than each of its constituent parts (SEAs Z, W and Y), and not less than all of its parts, is considered ivaemye separately. This "someone" can not be any of the subjects of SEAs Z, W, Y, because the vision of each of them is limited by the framework of the corresponding SEA. Therefore, "someone" should appear and form the integrating goal of the SEA S, its technology, etc. Perhaps, it will become a subject (STS) playing the role of a subject of one of the existing SEAs, or their composition. However, in any case, the necessary role of understanding the system as a whole (SEA S) is different from the roles of the subjects of the component parts of the system (SEAs Z, W and Y).

Thus, based on the postulate of "systemic" complex activity, composition and logical sequence of the procedural components of the activity, we can formulate the following statement.

The order of generation of complex activity elements

Generation of elements of complex activity is carried out in sequence only from the higher (according to the logical structure) to the lower ones, including in the case of self-organization.

The question of what the corresponding hierarchy should be is the subject of optimization.

All SEAs by duration of existence can be attributed to the groups of constantly existing (during the whole period of the activity review) and promptly generated, implemented and completed. The first, as a rule, are process ones, and the second ones are projected SEAs.

We introduce the classification of CA and SEA on a system-wide basis: the creation of new elements of CA, new SEA, taking into account the generally accepted division of any activity into *reproductive* and *productive*.

We distinguish four classes of activity and SEAs (Fig. 4.2).

(1) *"Regular"* CA and "regular" SEAs. Activity in the process of which new elements of activity, new SEAs, arise only because of deterministic decomposition of higher SEAs, respectively, a priori known technology (which is deterministic demand). The structure and technology of regular CA are deterministic.

(2) *"Replicative (passive)"* CA and "replicative (passive)" SEAs. Activity that result in activity of a known type (known needs, goals and technologies), and the non-trivial component of the activity is not to form, but to fix an undefined demand.

(3) *"Replicative (active)"* CA and "replicative (active)" SEAs. In the course of such activity, an undefined demand is formed, a new demand of a certain type (perhaps, a new consumer) is formed and, as a consequence, a new activity of a known type.

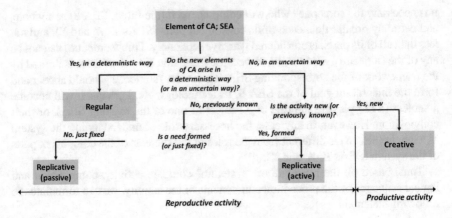

Fig. 4.2 Classification of CA and SEAs

(4) *"Creative"* CA and "creative" SEA activity, which result in an uncertain a priori demand for results of an unknown a priori activity, the technology of which must be created in the course of this new activity.

Regular, and both replicative activity correspond to *reproductive*, and creative— *productive activity*. Note that, depending on the objectives and level of consideration, the same activity can be interpreted as one or another in accordance with the classification shown in Fig. 4.2.

In the next Sects. (4.2–4.4), the features of the CA elements of all four classes will be discussed in detail, and the corresponding illustrative examples are given.

4.2 Regular Activity

One of the key trends of the current stage of human development is that any activity is purposefully subjected to maximum *simplification, standardization* and *regulation* as much as possible. This trend, let's call it the regularization of activity, as a mass phenomenon emerged in the STS during the first industrial revolution and the beginning of a serial factory production. Standardization and regulation of activity are one of the main ways to improve its effectiveness. Automation, in turn, continues this trend, replacing the activity with machine operations, displacing a person from the field of routine (highly regulated) operations and leading to the disappearance of many professions. During the last decades, the profession of telephone operator, typist, typographic workers and operators in many industries has thus disappeared. There is almost no doubt (a matter of time) in the relatively rapid mass replacement of drivers of motor vehicles with automatic weapons, etc. Information portals of state and municipal services that regulate citizens' access to government services and significantly improve the efficiency of their provision were widely disseminated.

Generally speaking, regular activity has always existed, and in the period of handicraft, and earlier. In past historical periods, the proportion of "regular activity" was less, since technologies have not been worked out. For one artisan-master (who created technology, that is, engaged in creative activity) accounted for a dozen apprentices (regular activity). Now, "all" employees of factory workshops (including complex assembly) and line departments of companies (including high-tech-logical) are engaged in "regular activity". The percentage of people engaged in non-regular activity is quite small.

Why a simplification (with the preservation of functionality), a regulation and a standardization improve the efficiency of activity? Because this reduces uncertainty, and this allows you to predict the results and plan activity on many steps forward.

Regularization of activity naturally has its drawbacks. Obviously, the *standard* (standardized, standard) *solutions* based on average conditions are worse than those based on an analysis of the specific specifics of each case. However, for the analysis and adoption of specific solutions, higher qualifications of subjects and substantially higher costs for their maintenance are needed. It also takes time and effort to identify each specific situation. And the time delays in making decisions also lead to losses. Therefore, the activity is regularized in those cases when it is (economically) justified.

Regularization of complex activity is one of the reasons for changing the role of information and information model of the subject and CA (Sect. 2.6). The value of the information model created in parallel with the realization of CA and the evolution of the subject increases. As a result, the influence of information technologies increases as a means of production of information models, the problems of knowledge management acquire special importance (see Sect. 2.6).

The following four are fairly vivid manifestations of the regularization of activity.

First, it is the popularity of various methods and models for optimizing production and business processes LEAN, CMMI, 6-sigma and other similar ones. The effect of regularization of activity is very significant, and this explains the spread of these methods. The technologies for performing operations, and the result, and the qualifications required for this are standardized. Many standard functions are automatized. "Perfect employee", not only in production, but also in the office, must strictly follow the regulations, achieve exactly the result, which is determined by the regulations, being in fact a "live automaton". In the early twentieth century, this approach began to be applied in mass conveyor production, then spread to the "operating" staff of large international corporations [53], and then to many other areas of activity.

Secondly, those who are interested in the results of complex activity (customers of complex technical and organizational and technical systems) require the use of only high-level technologies to reduce technological uncertainty. In this connection, the TRL-concept (TRL—*Technology Readiness Level*), which assumes the definition of formal maturity levels of technologies and the application at various stages of creation of a complex system or product of technologies below a certain level of maturity. Such concepts are developed and used by government agencies and large Western corporations, for example, the US Department of Defense [141], NASA [77], the US Department of Energy, Boeing, EADS and others. Moreover, highlighting "levels of complexity" is also characteristic of scientific activity—see, for example, the levels of complexity of inventive and cognitive tasks in [2].

Thirdly, the regularization of activity now concerns the activity of even top managers—the practice of "Corporate Governance" is intensively developing, which is nothing more than the establishment of certain regulations and standards for the activity of top managers of firms to protect the interests of shareholders, which is equivalent to striving for efficiency in terms of shareholders. Similarly, the activity of civil servants, including those in the highest positions, are limited and regulated to ensure the effectiveness of their activity from the point of view of citizens of persons interested in the results of their activity.

Fourth, the regularization of activity has also spread to "start-ups" innovative companies engaged in new developments, which are characterized by a very high level of uncertainty. Due to what regularization gives effect in such companies? The fact is that complex activity include a large number of elements, and not all elements are characterized by an equal level of uncertainty. Structuring and standardization of activity allows us to identify those elements, SEAs that are well known, and standardize them, exercising special control over the rest. Thus, it is possible to somewhat reduce the uncertainty and increase the efficiency of even such companies.

Thus, we can conclude that the proportion of regular activity is great and continues to grow. At the same time, the development of technologies extends the possibilities of automating routine, that is, regular, activity and these opportunities are realized, displacing entire professions (see examples in this section above).

Despite the "inseparability" of the person and activity, in the case of regular activity, the replacement of one subject (an individual) by another (with a given technology), or sometimes even an "automatic", does not change the results of the activity.

This conclusion to some extent corresponds to the forecasts of R. Kurzweil [72, 73] about the possible "*technological unemployment*" employment of people's jobs by robots.

Probably, all types of regular activity can be considered as potentially fully automated, and the chosen basis (deterministic technology and the generation of only known elements of activity only a priori) of the definition of regular activity can be a criterion for the separation of potentially automated and non-automated activity.

Regularization has two interrelated components:

1. Obtaining knowledge and reducing the uncertainty of CA in two ways: as practicing in practice a repetition of activity or as the study of fundamental laws.
2. Fixing knowledge in the form of an information model of activity, first of all, its technology.

In fact, the regularization of CA is a "knowledge of CA" fixation and systematization of knowledge about CA. From this point of view, regularization and similar actions can be considered as special cases of "scientific" activity. The following functions of science (and modeling as a method) stand out [98]:

- *Descriptive (phenomenological) function* is that due to abstraction of the model it is possible to explain quite simply the phenomena and processes observed in practice (in other words, they answer the question "how is the world arranged?"). Successful models in this respect become components of scientific theories and are an effective means of reflecting the content of the latter (therefore, the cognitive function of modeling can be considered as a component of the descriptive function).
- The *predictive function* of modeling reflects its ability to predict the future properties and states of the simulated systems, that is, to answer the question "what will happen?".
- The *normative function* of modeling is to get an answer to the question "how should it be?"—if, in addition to the state of the system, the criteria for estimating its state are set, then using the optimization it is possible not only to describe the existing system, but also to build its normative image—point of view of the subject whose interests and preferences are reflected by the criteria used. The normative function of modeling is closely related to the solution of control problems, that is, with the answer to the question "how to achieve the desired (state, properties of the system, etc.)?".

Those primary generalizations and systematization at the level of the describing function can then be used as "samples" at the level of the normative function.

Let us now consider the influence of measurable and true uncertainty on the regular activity in the context of the grouping of sources of uncertainty introduced in Sect. 2.7: the external environment, technology and subject.

Regular activity, regular SEA, are characterized by the fact that the actions are repeatedly repeated in accordance with effective technology. Repeatability of actions, in turn, means that the technology is tested with a wide variety of combinations of "normal" conditions (environment, technology factors and active behavior of the subject) in which this regular activity will be carried out. Therefore, in the presence of a sufficient number of used, consumed and accumulated resources (see Sect. 2.9), the results are in fact known (can be predicted with the required accuracy) already at the commencement of operations, the time of completion of actions and the amount of resources expended remains unknown.

If the conditions of activity do not differ from "normal", then the subject of the CA does not need a new goal-setting, structuring tasks, changing technology; the activity of the subject's behavior, his uncertainty is limited by the possibility of making a binary choice: "Should he act as a subject of a CA, that is, conscientiously perform the actions of a given technology, or refuse to participate in the CA?" In practical cases, this manifests itself in the form of dismissal of employees at their own will, absence from sickness or similar situations (facts of unfair performance with subsequent dismissal also fit into the above cases).

The technology of regular activity is stable, many times verified and does not require changes, so technological uncertainty can be caused only by events of inadequacy of means, methods, factors to CA tasks. In practice, such events are reduced to failures and failures of equipment, production of substandard shipments of raw materials, materials and components and similar cases.

The external environment does not affect either the goal setting or the technology of the regular CA, so the uncertainty of the external environment in this case is reduced to the possible emergence of conditions when either the technology turns out to be inadequate to the tasks of the activity or the subject will change his decision (on participation in the CA). A practical manifestation of the uncertainty of the external environment is the emergence of natural, economic, social, political conditions that impede activity.

Uncertainty of the subject of regular activity in the vast majority of cases is expedient to consider measurable, based on statistical patterns (for example, dismissal of employees of mass professions and their morbidity). Failures and failures of equipment, the facts of obtaining poor-quality raw materials and materials for regular CA technologies are well described by the laws of reliability theory, so the technological uncertainty of regular activity is also measurable. Due to the study and repeatability of a regular CA, the possibility of an a priori occurrence of unknown events in it is negligible, in most practical cases such events are considered to be force majeure and are taken out of consideration. Therefore, in this case, it is also expedient to consider the uncertainty as a measurable one.

Thus, we can say that the uncertainty of regular complex activity is measurable, that is, the indeterminate events of all three groups (active subject choice, technologies and external environment) are statistically measurable.

In the case of realization of uncertainty, regular activity subjects face a situation that is beyond the scope of regulations and standards, their areas of responsibility, therefore they escalate problems to the higher level of the SEAs in the logical structure (see Sect. 3.3). "Non-standard" for the subjects of the situation are resolved at the higher levels of the hierarchy, while the regular CSA remains deterministic and does not generate a priori unknown elements of the CA.

Three important system-wide properties of regular SEAs, their structural models, follow from the determinism of technology (Fig. 4.3).

First, in the logical structure, all subgoals are mandatory for achievement. From this property follows the second, consisting in the fact that in the causal structure all

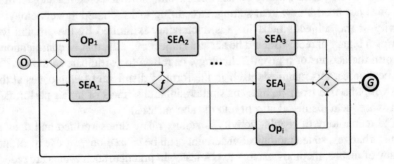

Fig. 4.3 An example of a regular SEA causal structure

the conditions for achieving goals that are a combination of several parallel goals are strictly conjunctive. Also from the first property follows the third property: in the causal model of regular activity there are only initial and terminal events, there are no intermediate events. Uncertainty events, which, however, always take place, are described outside the causal model (in the process model, which will be introduced in Chap. 5).

Examples of regular activity and regular SEAs are:

- "Preparation of daily reports on closing the operational day of a retail bank by accounting staff";
- "Making an order for certain components by an employee of the logistics department of an aircraft building company";
- "Fire safety calculation of the deployment of fire extinguishing equipment in the course of fire and technical exercises";
- "Installation of the pump of a particular power unit system during routine maintenance".

All these elements of activity are repeatedly repeated and detailed regulated.

4.3 Replicative Activity

Replicative activity (and replicative SEAs) is defined above as an activity in the course of which a passive activity is recorded or a demand of a certain type is created (active activity) (possibly, a new consumer of the performance results). As a result, a certain demand is actualized and a new activity of a known type is generated, creating known results. Replicative activity consists in revealing not new types, kinds, forms of needs, and new consumers of known results, that is it is activity on initiation of other activity.

Demand in the case of replicative activity is of an uncertain nature and is the main factor affecting it. We will call the uncertainty of demand the basic uncertainty of replicative activity.

If the demand has the character of a mass, typical, repetitive phenomenon, it does not need to be created, but only to be fixed, as it happens for passive replicative activity, then the basic uncertainty of activity is measurable. In the case of the creation or formation of demand (active replicative CA), it can not be considered typical and repetitive, at least at the initial stage. Therefore, the uncertainty in this case is true, and the basic uncertainty of active replicative activity is true.

Illustrative examples of passive replicative activity are:

1. A dispatcher or call center operator who receives messages, requests, orders and orders (fire services);
2. The client manager or the seller in the store, meeting the incoming customers and providing them services or selling goods (in the branch of the retail bank);

3. An employee or a working group performing diagnostic work of any equipment or controlling the quality of raw materials, materials, products (equipment of nuclear power plants or products of an aircraft construction company).

Active replicative is, for example, entrepreneurial activity (usually top-managers, vice-presidents), which has the character of replication, business development:

4. Employees of the aircraft construction company selling aircraft to airlines;
5. Employees of the retail bank on the analysis of opportunities and decisions on the opening of a new business of a known type—new branches of a retail bank, likewise—for the construction of a new workshop or plant to assemble an already produced model of an aircraft or its assemblies.

Let us consider the structures of the SEAs corresponding to these examples.

The activity in examples 1 and 2 (its structures are shown in Figs. 4.4 and 4.5) can be conditionally called the *dispatching activity*, which is reduced to:

- waiting for the next request or customer (SEA_1 or $operation_1$)—an undefined event as a manifestation of demand;
- responding to heterogeneous events ($u_2, u_3, \ldots, u_J$) in one of several ways ($G_2, \ldots, G_J$).

The logical structure (Fig. 4.4) reflects exactly the composition of the elements of the SEA.

The activity is realized in time (cause-effect structure, see Fig. 4.5) as follows: at the beginning of the Gr action, the "basic" initial vertex 0 is initiated, followed by

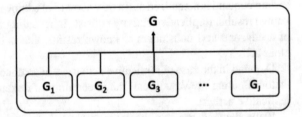

Fig. 4.4 An example of a logical structure of dispatching (passive replicative) activity

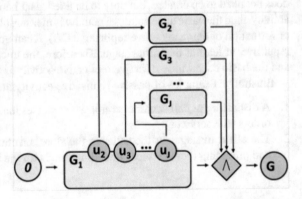

Fig. 4.5 An example of a causal structure of dispatching (passive replicative) activity

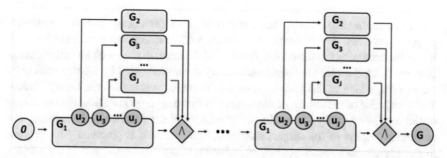

Fig. 4.6 A variant of a causal structure of corrective (passive replicative) activity

the SEAs G_1. In the event of any of the uncertain demand events, $u_2, u_3, \ldots, u_J$, one of the SEAs $G_2, \ldots, G_J$ is realized, and the expectation (G_1) is continued in parallel.

The diagnostic activity in the framework of Example 3 is reduced to the execution of one or several consecutive control operations, the results of which are organized elements of corrective activity. Such activity may have structures that coincide with the ones considered above (dispatching activity), or others, for example, presented in Fig. 4.6 the consecutive composition of the structures shown in Fig. 4.5.

The active replicative activity of Examples 4 and 5 can also be represented by the structures of Figs. 4.4 and 4.5, it is natural that inferior SEAs in these cases are much more complex.

Thus, it can be concluded that the causal structure presented in Fig. 4.5, has a certain generality and can be considered the basis for replicative activity.

When analyzing the examples, there were no restrictions on the lower SEAs, so there is every reason to believe that downstream SEAs of replicative SEAs can be process or project, regular or replicative, elements of replicative SEA can also be elementary operations.

Let us now consider the factor of activity, the active choice of the subject of replicative activity.

Passive replicative activity, like a regular one, does not require anything from the subject except conscientious fulfillment of actions, and the degree of conscientiousness is reliably determined by the consumer of the results—the subject of the superior SEA. The subjects of such activity do not influence the demand, that is, they do not affect the basic uncertainty of the activity. They do not reveal the demand and consumers, but only fix them. For example, customers themselves come to the bank branch, the task of the bank's employee is to fix this fact, determine the service that is necessary for the client and the provision of this service. The number of clients served as a result of the activity depends on the subject only to the extent of the resource constraint and the proper performance of its functions, to a much greater extent it is given by an undetermined stream of customers-demand independent of the subject. Uncertainty of the subject, as in the case of regular activity, is limited to his binary choice, whether to participate in the activity.

Active replicative activity has the character of promoting complex products, projects, services, development or replication of business. As it was determined, it is connected with the formation, creation of demand, that is, with an active choice of the subject, therefore the basic uncertainty is supplemented by the uncertainty of the subject, which in this case is substantially more diverse than the binary choice. The variability of the choice of the subject in this case does not allow the consumer of the result—to verify the results of the activity of the superior SEA, to verify the conscientiousness of the subject's actions. In this case it is practically impossible to reliably separate the dishonesty or imitation of activity on the part of the subject from the realization of the basic uncertainty of the activity.

In concluding this section, we note that replicative SEAs, in addition to their basic uncertainty, are also characterized by a measurable uncertainty, analogous to the uncertainty of a regular CA, described in Sect. 4.2. Unconventional situations caused by this measurable uncertainty can both be "resolved" at the level of the replicated SEA (within the framework of which they arose), and be escalated to the level of the superior SEA.

Important system-wide properties of replicative SEAs, the properties of their structural models, are the following. First, in the logical structure, only a part of the goals are mandatory for achievement. Secondly, the causal structure contains several undefined events and the corresponding chains following them. Thirdly, all downstream SEAs can be either regular or replicative.

4.4 Creative Activity

Creative activity (creative SEAs) is an activity whose technology is not fully defined (not fully known) at the time of the commencement of activity and is therefore created in the course of the realization of the activity. The uncertainty of the technology is caused by the uncertainty of demand and/or the a priori uncertainty in the specification of the result of the activity. This is an activity to obtain a result that is not fully specified at the beginning of the activity. This class includes activity of chief designers and technologists, researchers, producers of films and performances, partners of law firms, etc. Subjects of creative CA independently determine the structure and characteristics of a complex result and, consequently, the structure and technology of activity. In fact, they are the *architects of activity* (as a system, for a system architecture, see Sect. 2.10) and the architects of the result (as a system). The principal difference between creative CA from replicative and regular CA is the presence in the structure of the first, at least one fragment, the subject of which is the technology of another ("inferior") fragment of CA, which is a consequence of the need to create new technology in the performance framework.

It can be noted again that the maximum breakthroughs in any areas of human activity is given by creative activity, but it is not mass, and its successful realization is rather an art than is defined and described by science and technology.

The creation of new products, new system-architectural solutions and new technologies generally refers to heuristic, inventive, creative activity. Its key elements are elementary operations carried out by one individual, although the support of a complex subject can be provided by large working groups, which, as a rule, carry out regular activity. Complexation of the results of heuristic operations and regular fragments is performed just within the framework of *creative SEAs*.

For example, the chief designer of the aircraft takes the basic layout, architectural solutions, while modeling, calculations, structural workings of units and aggregates of the aircraft are performed by numerous engineering groups of employees of the aircraft construction company.

The presence of heuristic elements of activity leads to the fact that the key feature of creative activity is true uncertainty, immeasurable and "non-statistical." Moreover, since the subject of creativity is necessarily the technology being created, this uncertainty is "technological". At the same time, *heuristic activity* is highly subjective, so technological uncertainty is accompanied by uncertainty of the subject's actions. Naturally, it is not possible to reliably divide the influence on the result of the CA of technological uncertainty and the uncertainty of the heuristics of the subject, so it is advisable to consider them together. In many cases, these sources are supplemented by the uncertainty of demand and the external environment as a whole, forming a complex composition of sources of uncertainty. That is, the basic uncertainty of creative activity is true and is caused (not always shared) by the composition of non-determinism of the external environment, technology and subject.

Despite the complexity of the composition of the sources of the basic uncertainty of creative activity, based on system-wide considerations, it is possible to distinguish binary features associated with the elements of the SEA, and to structure the sources based on these characteristics. Let us make this binary structuring of the uncertainty of creative activity presented in Fig. 4.7.

After the subject actualized the need, carried out goal-setting and structured tasks, determined the shape of the future result and, consequently, the need to create a new technology. Therefore, the structuring binary features will be the following:

(a) Are there, or will technologies be created, or available, in order to get the planned shape of the result?
(b) Will the result have planned properties/functions?

Both of these issues relate to technological uncertainty, and the first—affects also the uncertainty of the heuristics of the subject.

The result, its properties and functions are needed not by themselves, but in interaction with the external environment, in response to demand, so there is the following group of structuring signs-questions:

(c) Will the resulting properties/functions of the result provide the desired effect in interaction with the external environment?
(d) Will the properties of the result correspond to the requirements of demand at the future moment of presenting the result to the consumer?
(e) Will the external environment at the moment of presenting the result correspond to the current forecast of its development?

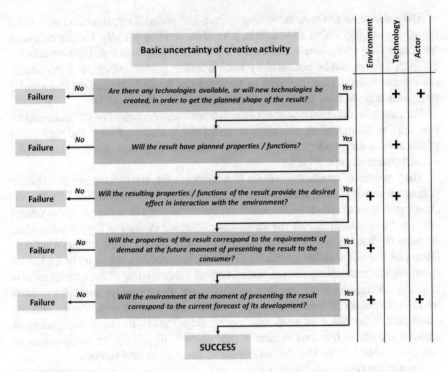

Fig. 4.7 The binary structuring of the uncertainty of creative activity

All three questions (c)—(q) are related to the uncertainty of the external environment, question (c) is also addressed by technological uncertainty, and question (e) is the uncertainty of the subject.

The table in Fig. 4.7 illustrates the origin of sources of uncertainty: the external environment, the technology, the subject.

As in the case of replicative activity, creative SEAs are also characterized by a measurable uncertainty, analogous to the uncertainty of a regular CA (except for their basic uncertainty). Unconventional situations caused by this measurable uncertainty can both "resolve" at the level of creative SEA, and escalate to a higher level.

The structural properties of creative SEAs substantially coincide with the properties of replicative ones, which is not unexpected, since both classes of SEAs generate activity. The principal differences are the a priori uncertainty of the creative CA technology and the complexity of sources of its basic uncertainty against the known technology and uncertainty caused by demand only, in the case of replicative. The a priori unknownness of technology is expressed in the presence in the structure of the creative SEA of specific SEAs reflecting the creation of a priori unknown technology. The creation of a new technology is a special case of complex activity; therefore it is adequate to describe it in the form of SEAs. A system-wide model of the process for creating a new technology is presented in Sect. 5.4. In the cause-effect structure of creative CA such SEAs are in chains following the uncertainty events (u_1 and u_2 in Fig. 3.9).

Just as for active replicative activity, the result consumer—the subject of the higher SEA—is not able to reliably verify the results of activity, dishonesty or imitation of activity on the part of the subject from the implementation of the basic uncertainty of the activity.

Examples of creative SEAs are the following. For an aircraft construction company it is the activity of the chief designer of a product (aircraft) to design the overall layout and overall appearance of the product. In many cases, its solutions are heuristic, although they are based on numerous calculations and justifications. Similarly, the decisions of the vice-president and/or the relevant committee deciding whether to approve the requirements (primarily operational ones) for the designed product. In each of these cases, the whole complex of sources of uncertainty is present, corresponding to the structure of Fig. 4.7.

In the case of a retail bank, the design of a new banking service (an analogue of the product design) or the approval of the requirements for a new service project is described.

4.5 The Role of Uncertainty

Complex activity, its elements, SEAs exist in time in the form of life cycles (Fig. 2.1), representing a set of ordered time intervals of phases and stages. Under the *implementation of uncertainty*, we will understand the occurrence of an event/the fact of the receipt of new information on the influence of undetermined factors on the elements of a CA. The uncertainty of the CA (Sect. 2.7) is realized in the form of unpredictable (or predictable, but random) events that affect the performance of the SEA and its result. These events can have a different nature and have a different effect on CA. So, for example, uncertainty can "increase"—for example, if an event is realized that consists in the completely unpredictable influence of external factors on the parameters of one or another SEA. Or, on the contrary, uncertainty can "decrease"—for example, if the exact value of a previously undefined parameter has become known (examples from decision trees or games in expanded form—the implementation of an indeterminate parameter: the course of nature or other subjects).

The combination of CA implementation and realization of uncertainty in the most general case leads to *the model of the SEA state dynamics*, the transitions between which are initiated by the events of the SEA, the uncertainty and the life cycle of the CA (see Fig. 4.8).

Regardless of the specifics of the activity, within the framework of the SEA, activity may or may not be performed, there may occur cases of measurable and immeasurable uncertainty.

Then, at the level of abstraction from the specifics of each specific activity, the possible states of any SEA can be represented by the following list:

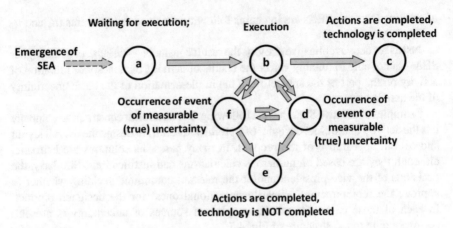

Fig. 4.8 SEA execution: states and transitions

a. SEA was created, but activity is not started—waiting for the beginning of the life cycle.
b. Implementation of the stages of the LC.
c. All the actions provided by the technology are completed, the result is obtained.
d. Reaction to an event of measurable uncertainty that occurred earlier.
e. The activity is terminated; the actions provided by the technology are not completed.
f. Reaction to the event of true uncertainty, which occurred earlier.

All possible transitions are reflected in the diagram of Fig. 4.8.

The state *a* is the initial one, corresponds to the time interval when the SEA subject decided to perform the activity, but the activity has not yet been started. The state *b* corresponds to the successive execution of all stages of the SEA's CC. When an event of measurable (or true) uncertainty occurs, a transition to the *d*—arrow from *b* to *d* occurs (transition to the state *f* from *b* to *f*)—a reaction to the uncertainty event occurs. As a result of the reaction to the uncertainty, a return to a routine performance (arrows from *d* or *f* to *b*) may occur, or there may be another event of uncertainty on which the reaction will be performed (arrows from *d* to *f* or from *f* to *d*), or will be accepted the decision that a regular continuation of activity is impossible and it must be stopped (arrows from *d* to *e* or from *f* to *e*). In the event that during the CA realization the events of uncertainty did not occur, all the actions provided by the technology are successfully completed—the arrow from *b* to *c*. States *c* and *e* are terminal, the transition to them actually means the continuation of the SEA in the form of historical data, and the result of the CA, of course.

The fixing of events of uncertainty is carried out by the subject, both directly and indirectly, as a result of escalating problems by subjects of lower-level SEAs.

The response to events of uncertainty implementation may result in the generation of new SEAs or in the transmission of information on the escalation of the problem to the higher SEA.

The model of uncertainty implementation is a generalizing model that represents all the classes identified above—regular, replicative and creative CA. This model in the form of a state diagram and possible transitions between them is methodically close to the formalisms of cellular automata and discrete event system (Discrete Event System Specification) [96], widely used for modeling complex systems. Together with the SEA formalism and the structural models cited above, the model for the realization of uncertainty forms a set of models representing various aspects of complex activity. In the next chapter, models of life cycles of CA, integrating structural models and models of activity generation, will be presented.

In conclusion of this chapter, it makes sense to return to the trend of regularization of activity, identified in detail and discussed in detail above (Sect. 4.2), which in fact consists in the "cognition of complex activity" of fixing and systematizing knowledge about it. In fact, this work and Chap.3 in particular is an example of regularization: the identification and fixing of system-wide regularities of complex activity is nothing but the first step in the regulation of activity. Moreover, since all classes of CAs are considered, regulation is also subject to creative activity, which is distinguished by an exceptionally high level of a priori uncertainty.

Summary of This Chapter

Classification of complex activity types and its elements on a system-wide basis is introduced: the generation of new elements of CA (new SEA); four classes are distinguished:

(1) *"Regular"* CA and "regular" SEAs. Activity in the process of which new elements of activity, new SEAs arise only as a result of deterministic decomposition of higher SEAs, respectively, a priori known technology (which is deterministic demand). The structure and technology of regular CA are deterministic.

(2) *"Replicative (passive)"* CA and "replicative (passive)" SEAs. Activity that result in activity of a known type (known needs, goals and technologies), and the non-trivial component of the activity is not to form, but to fix an undefined demand.

(3) *"Replicative (active)"* CA and "replicative (active)" SEAs. In the course of the activity, an undefined demand is formed, a new demand of a certain type (perhaps, a new consumer) is formed and, as a consequence, a new activity of a known type.

(4) *"Creative"* CA and "creative" SEAs. Activity, which results in an uncertain a priori demand for results of an unknown a priori activity, the technology of which must be created in the course of this new activity.

The analysis of properties and features of each class is carried out.

Regular activity. The system-wide properties of SEAs and their structural models follow from the determinism of technology. First, in the logical structure, all subgoals are mandatory for achievement. From this property follows the second, consisting in the fact that in the causal structure all the conditions for achieving goals that are a combination of several parallel goals are strictly conjunctive. Also from the first property follows the third property: in the causal model of regular activity there are only initial and terminal events, there are no intermediate events. Events of uncertainty, which, however, always take place, are described outside the causal model. Demand is deterministic, it reflects a logical structure, and uncertainty is measurable. Sources of measurable uncertainty are the external environment, technology and subject and subject matter. Uncertainty of the subject in the case of regular activity is limited by his binary choice, whether to participate in the activity. In the case of realization of uncertainty, the subjects of regular activity face a situation that goes beyond the limits of regulations and standards, their areas of responsibility; therefore they escalate the problems to the higher in the logical structure of the level of SEAs.

Replicative activity. The important system-wide properties of replicative SEAs, the properties of their structural models, are the following: First, in the logical structure, only a part of the goals are mandatory for achievement. Secondly, the causal structure contains several undefined events and the corresponding chains following them. Thirdly, all downstream SEAs can be either regular or replicative. Uncertainty of demand is the main uncertainty. Replicative SEAs, in addition to their basic uncertainty, are also characterized by measurable uncertainty, analogous to the uncertainty of a regular CA. Non-standard situations caused by this measurable uncertainty can both be "resolved" at the SEA level within which they originated, and be escalated to the level of the superior SEA. In the case of passive replicative activity, the subject's uncertainty (as in the case of regular activity) is limited to his binary choice, whether to participate in the activity. In the case of active replicative activity, the variation of the subject's choice is very high and has a very significant effect on the result. This does not allow the consumer of the result—to the subject of the higher SEA to verify the results of the activity, to check the integrity of the subject's actions.

Creative activity. Properties of structural models of creative SEAs almost coincide with the properties of replicative SEAs. The key property of creative CA is that the basic uncertainty is true and is caused (not always shared) by the composition of the uncertainties of the external environment, technology and subject. The a priori unknownness of technology is expressed in the presence in the structure of the creative SEA of specific SEAs reflecting the creation of a priori unknown technology. To structure the basic uncertainty, a system of binary features is proposed (Fig. 4.7). Just as for the active replicative activity, the result consumer—the subject of the higher SEA—is not able to reliably verify the results of creative activity, dishonesty or imitation of activity on the part of the subject of creative CA from the realization of the basic uncertainty of the CA.

A model was proposed for the implementation of uncertainty in the course of performing a complex activity the model of the SEA states, the transitions between which are initiated by the events of the CSE, the uncertainty and the life cycle of the CA (Fig. 4.8).

In conclusion, we give a comparative system-wide properties of regular, replicative and creative SEA (see Table 4.1).

In Table 4.1 the following designations are used: sources of measurable uncertainty, the external environment, technology, subject and subject matter are inherent

Table 4.1 System-wide properties of SEAs of different classes

Class of activity	Technology	Source of uncertainty	Activity of subject
Regular	Known, is not created during CA	Sources of measurable uncertainty	Binary choice
Passive replicative			
Active replicative			Active and binary choice
Creative	A priori unknown, is created during CA	Environment, technologies, subject, sources of measurable uncertainty and demand	

in all classes of CA; binary choice of the subject is inherent in all classes of CA; the active choice of the subject in the case of replicative CA is aimed at the formation of demand, in the case of creative—on the creation of technology and on the forecasting of demand and other factors of the external environment.

A practically important consequence of the introduction of activity classes is the definition of formal characteristics that allow us to separate the elements of activity based on their inherent uncertainty. The possibility of justifying a relatively small number of elements that are inherent in true uncertainty, among the mass of regular elements, whose uncertainty is measurable and small, makes it possible to address more appropriately and, accordingly, more effectively the practical tasks of risk management and those adjacent to them.

The list of requirements to be met by the methodology of complex activity and the system of CA models was formulated in Sect. 2.10. The classification of SEAs and models of the generation of activity and the implementation of uncertainty ensure that the requirements in section (e)—(g) are fulfilled. An essential requirement for the methodology of complex activity is the need to describe the life cycles of CAs, the development in time, the realization of CA elements in the general case, taking into account the uncertainty, so in the next chapter we will consider models that meet the requirements of (e) … (h).

Chapter 5
Process Models of Complex Activity

The realization of complex activity occurs in the general case in the form of generation, realization and evolution of a set of interacting elements of a complex activity that form a hierarchical structure (for example, see Fig. 2.4). In the previous chapters, structural models of CA have been described and the generation of CA elements has been analyzed. Being generated, the CA elements are realized and evolve in time, so we will now consider the process models of the SEAs of the model for realizing their life cycles, the models representing the evolution of CA elements, their behavior over time.

The life cycle of the CA element reflects cause-effect relationships, so let us explain the need for introducing process models along with causal models of CA elements. The causal model represents the same links between the goals of the subordinate level, these links are specified in technology and are manifested at the stage "Carrying out actions and obtaining a result" of the life cycle of the CA (Fig. 2.1). The causal model does not describe the processes occurring during the design phase, nor during the phase of reflection (Fig. 2.1), as well as the occurrence of uncertainty events and the reaction to them. The necessity of modeling these processes makes it important to use models for the execution of the LC of CA—process models, which this chapter is devoted to.

The development of any model involves the isolation of features, characteristics, properties of the phenomenon or object being studied and their reflection with the help of some description tool. In this chapter, the emphasis is on system-wide features of the *procedural components* of complex activity, the processes of *realizing their life cycles*. Therefore, we use the phases and stages of the LC of the CA (as defined above—see Fig. 2.1) as a basis, and we will analyze and detail them for constructing models.

Representation of models will be performed in BPMN notations [22].

In the framework of this chapter, the analyzed and modeled activity will be called "A-SEA", as well as an *agent*; the superior (upper) in the logical structure of the SEAs will be called the *U-SEA* (upper), and the lower-level elements (SEAs and operations) will be called *L-SEAs* (lower) and *L-Op*.

M. V. Belov and D. A. Novikov, *Methodology of Complex Activity*, Studies in Systems, Decision and Control 300, https://doi.org/10.1007/978-3-030-48610-5_5

The structure of the presentation of material in this chapter corresponds to the logic of complicating the life cycles of CA elements.

The simplest case is the LC of an elementary operation, but this life cycle is degenerate: it is not detailed and is presented in the form of a single stage. Its system-wide process model is trivial consisting of a single specific object and does not require further description, and the formal expression of the process model of an elementary operation coincides with its cause-effect structure presented in Fig. 3.10 in Sect. 3.4. According to the principle of generalization and abstraction (Sect. 1.3), each element of a CA can be considered both as an elementary operation and as SEAs, depending on the researcher's goals, which entails an aggregated or expanded representation of the element to models. This fully applies to the process model: if the consideration of phases, stages and stages of the life cycle of a CA element is necessary from the point of view of the researcher, then regardless of the simplicity or complexity of the logical and causal structures of this element, it must be represented as SEA, and for him a full-fledged process model should be formed.

The first section is devoted to the analysis and modeling of the essential and typical component of the life cycles of all CA elements of the process of demand formation and actualization of it as a need. In the second section, a process model of the structural element of the SEA CS is presented, as a basic element of complex activity. In the third and fourth sections, system-wide models of typical elements of activity were introduced: the creation of a new CA technology, which is an indispensable stage in the generation of a new CA, and supporting activity for organizing resources. The final (fifth) section of this chapter is devoted to modeling the process of performing complex activity as a system.

5.1   Models of Demand Generation and Actualization of Needs

Demand formation and actualization of the relevant need by the potential entity is a standard procedure (conditionally call it the *procedure of demand generating—PDG*, the demand for performance results), which is repeatedly performed within the elements of the CA. From it begins the execution of the LC of any activity, so it makes sense to consider it in more detail, highlight system-wide properties to reflect them in the model.

First, the result of the DG is that the individual (or STS) is aware of himself as the subject of activity, in fact, the self-creation of the subject of the CA element occurs. That is, in PDG there is always at least one participant—a future subject with the ability/property of active choice. We will find out if only one participant of the procedure can always be active, and for this we will analyze again the sources of demand (Fig. 5.1).

In Sect. 4.1 it was shown that demand can be caused by one of two factors: the decomposition of an already existing CSA and the occurrence of uncertain events. In

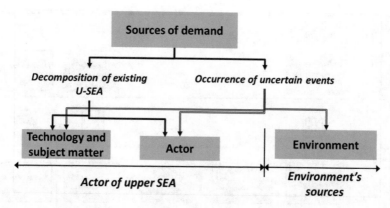

Fig. 5.1 Sources of demand

Sect. 3.2 (Table 1.7), it was illustrated that the decomposition of SEA can be carried out by subject, technology and subject. The analysis performed earlier in Sect. 2.8 allowed us to identify three groups of sources of indeterminate events:

(a) the external environment,
(b) technology and subject matter,
(c) the subject.

The union of all classifications allows us to formulate the following statement.

> **Sources of demand for performance results.**
>
> Sources of demand for the results of a new element of complex activity can be only the subject of the superior element of the CA or environmental factors.

Left or right side Fig. 5.1 (and, accordingly, the cases of 1 or 2 initiators in Fig. 5.2) illustrate this statement.

That is, in a number of cases, the PDG is performed by two participants: the subject of the U-SEA, who is the initiator, and the individual (or STS), resulting in the subject A-SEA or A-Op, each subject having an active choice. The presence of two active participants requires consideration of the PDG from the point of view of each of them, therefore the model (Fig. 5.2) is represented in the form of two parallel "paths", each of which contains the steps performed by the relevant subject.

In private, but not very rare, the case of the role of both actors can play the same individual (or STS), acting as the subject of two elements of activity. One of the examples discussed in this section below (about the employee of the branch of a retail bank serving customers), corresponds exactly to this case.

Secondly, procedurally, the formation of demand and its actualization take place in the form of transfer of information from the initiator to the individual or the organizational system (block 1 in Fig. 5.2), which, having received information

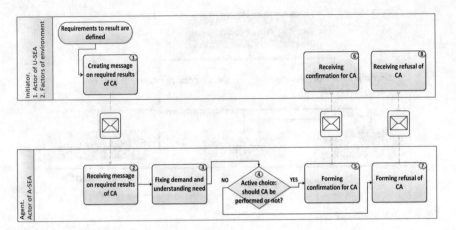

Fig. 5.2 PDG and turning it into its internal need

(block 2), actualizing demand and turning it into its internal need 3), become the subject of the child element of the CA—SEA (block 4).

After that, the new entity, as a rule, passes to the initiator the confirming information that the demand is updated (block 5), and begins the implementation of the business center or the refusal (block 7).

As a result of the described actions, the connection between the initiator and the subject actually occurs, that is, the organization appears as a property; and the corresponding process is the process of organization.

Demand reflects the specific requirements for the expected performance of the agent, so the transfer of information can be represented as a message exchange with a known a priori set of attributes—the exchange of pieces of information that occurs at some points in time. A particular case, when information is transmitted and received in time continuously, is reduced to the above: since the individual (the potential subject of the new activity) can decide whether to perform the activity, only after receiving all the information, we can say that he receives the message at the time of receiving the final portions of information.

In practice, information exchange in the overwhelming majority of cases occurs as an order/appointment and confirmation (but can be carried out in other forms, in particular, as simple information). For example, as an instruction to an employee (or working group, project team, department, etc.) to perform work (activity) and appropriate confirmation.

Thirdly, the PDG is of an uncertain nature, we will consider sources of uncertainty from the point of view of each of the participants.

From the point of view of an agent resource, the "potential" A-SEA entity, the PDG consists in deciding whether to participate in this activity as its subject, and the uncertainty lies in the presence or absence of initiative on the part of the initiator. The uncertainty of the initiative is characterized by any of the sources presented in Fig. 5.1: technology and subject, subject, environment.

If the initiator of the PDG is the subject of the U-SEA, then for him the uncertainty is manifested during the procedure: for some reason there may not be an individual (STS) who can or will want to perform the required CA element, therefore, in general, the subject of the child CA element can not be formed. For the initiator, the uncertainty of the PDG is caused by the "binary choice" of the agent (the concept of binary choice is introduced and discussed in Chap. 4).

The direct organization of the PDG is specific, but its essential aspects are:

1. presence of one or two participants with the possibility of active choice;
2. external (in relation to the element of activity, the demand for which is formed) the nature of demand and the formation of its external environment or superior SEA;
3. execution of the procedure in the form of an exchange of messages a priori of a given structure;
4. execution of the procedure within a certain period of time (if it is expected to receive a response message from the transaction subject) or one-time (if no confirmation message is required);
5. uncertainty (from the point of view of the agent, the uncertainty of the environment, from the point of view of the initiator, the uncertainty of the agent's active choice).

These aspects are invariant to the subject area of activity, and they are reflected in the model, the block diagram of which is presented in Fig. 5.2.

A significant special case is the procedure for generating demand for the results of activity on the organization and management of the life cycle of resources (Sect. 2.9), we will conditionally call it the *procedure for requesting, receiving and organizing resources* (PRROR). The PRROR is performed (as a rule, repeatedly) during the execution of each element of the CA: in the formation of the subject, technology and subject. This procedure is often used in the models of this section, so it makes sense to bring its block diagram (Fig. 5.3), formed on the basis of the PDG block diagram.

The application of this model of PDG can be illustrated by the following examples.

In the case of a retail bank, a client coming to a bank branch causes a demand for the result of an activity that involves providing him with certain services. Modern banks use several options for implementing customer service procedures. The traditional option, used since the beginning of the development of the market of mass banking services, is the organization of several workplaces in the service hall, on each of which the bank employee provides a set of services to clients.

The activity of employees at these workplaces is "control room" (i.e., passive replicative activity, see examples in Sect. 4.3, Figs. 4.4 and 4.5). The client, having come in branch of bank, chooses, to which of employees to address. Such employee, within the framework of his dispatching activity, determines what kind of services a client wants to receive, and thus creates a demand for (his) activity to provide the necessary service. The exchange of messages in this case is degenerate, as it happens "in the head" of the employee of the bank, which at that moment becomes the subject of the element of activity for the provision of a specific service. In a more complicated variation of the procedure, the bank appoints a hall manager who meets the client,

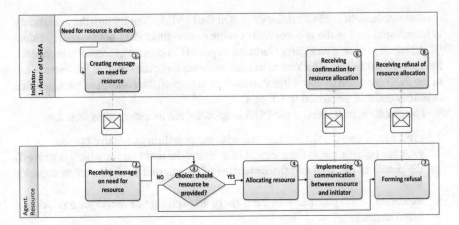

Fig. 5.3 The procedure for requesting, receiving and organizing resources (PRROR)

finds out from him what service is needed and directs him to a free employee engaged in the provision of relevant services. In this case, the hall manager is not responsible for the servicing staff. Therefore, it is expedient to consider its activity not as a replicative activity that generates demand for the activity of serving employees, but as the routing of the flow of customers and the creation of a "friendly" and "client-oriented" image of the bank. "High-tech" (and, probably, economically advantageous in the long term) option is to replace the hall manager with an electronic self-service kiosk that creates an "electronic queue". Often the bank, together with the installation of the kiosk, has to maintain a room manager who helps clients who do not have the skills and sufficient level of "computer literacy" that allows them to use the kiosk on their own. It is important that the activity of the serving employees remain the same as in the first case.

Quite the same, there is a formation of demand for the activity of firefighting in the fire department. The operator of the unified municipal "center 112" (emergency control center) receives messages and transmits them to the operative on duty of the fire department, which, within its "dispatching" replicative activity, generates demand for firefighting activity, transmitting the corresponding message to the commander of the calculation.

Similarly, the PDG is implemented in the example of the nuclear power plant, discussed in Sect. 4.3 (Example 3, Figs. 4.4 and 4.5).

In another example (Example 4 of Sect. 4.3), the vice-president of an aircraft construction company, having concluded a contract for the sale of a series of aircraft, confirms or corrects the preliminary specification of an order already available in the firm's information systems, and thereby generates a message (informationally rich and complex, but, nevertheless, formalized) to a group of employees performing production planning. This group actualizes the demand thus expressed and starts a complex activity (including, perhaps, tens or hundreds of thousands of SEAs and a few years) for the production of ordered products.

5.2 Process Model for Elements of Complex Activity

Analysis and modeling of the system-wide features of the life cycle of the SEA will be carried out based on the properties of SEAs and described above of structural models (Chap. 2), models of generation of CA elements and uncertainties (Chap. 3), models of demand generation (Sect. 5.1).

The realization of the CA element includes all phases and stages of its life cycle introduced above (see Fig. 5.8), during which elements of organizational and managerial activity are also realized. These phases and stages will be reflected in the process model in the form of a sequence of objects, which we conventionally call the "main chain", since they represent the activity itself. Uncertainty is inherent to any CA, so the process model should also contain objects that reflect the onset and detection of events of uncertainty and the response to them, call them a *"chain of reaction on uncertainty."*

Depending on the CA class, the characteristics of the two main chains and the response to uncertainty have their own specificity; we describe it using the results of Chap. 4.

Regular CA does not allow the creation of new elements of activity. This means, on the one hand, that all L-SEAs and L-Ops must be performed, and on the other hand, the unambiguity of the response to uncertainty in the form of an escalation of the problem to the level of the superior SEA (Sect. 4.2) is unambiguous. Therefore, the "main chain" of the LC of a regular SEA includes the execution of all L-SEAs and L-Ops (represented in both the logical and causal structures), and the "chain of response to uncertainty", as a reaction to uncertainty events—escalation of the problem to higher SEA.

Replicative activity generates a priori known elements of CA. Only SEAs following the initial event (marked "0") in the cause-effect structure (for example, SEA_1 in Fig. 4.5 of Sect. 4.3) are mandatory for execution, all these SEAs will be executed in the "main chain". In contrast, the implementation of SEAs following events of uncertainty (for example, SEA_2, following the event u_2 in Fig. 4.5), is a response to uncertainty. Therefore, the onset of uncertainty events and the implementation of the following SEAs, together with the possibility of escalating problems, is the content of the "chain of reaction to uncertainty".

In the case of creative CA, the content of the "main chain" and the "chain of response to uncertainty" at the level of the structures will be completely analogous to the regular CA. The difference lies, as noted in Sect. 4.4, in the availability among SEAs of those that represent the creation of new technologies (the model of this SEA is given below in Sect. 5.3).

In view of the prevalence of elements of regular activity (see Sect. 4.2), we shall separately consider the case of LC of such SEAs. Recall that regular SEAs are characterized by a priori known and repetitive demand, the structure of goals and technology: at the time of fixing demand, the logical and causal models of the element of activity already exist, as well as the required resource pools for their implementation.

Let's consider the stages of the implementation of the SEA LC, starting with the "main chain", we illustrate them by Table 5.1 and the BPMN model, shown in Fig. 5.4.

The realization of the CA of any SEA is in itself a complex activity: phases, stages and stages are elements of activity. From this point of view, the activity on the implementation of the LC is an SEA consisting of lower-level operations and SEAs, and the process model is the cause-and-effect model of the SEA "Implementation of the SEA CA"; we use BPMN notation to represent the model. Elements of the BPMN-model (elementary operations and SEAs) will be numbered according to the numbers of the stages of the LC in Table 5.1.

The implementation of the CA of any SEA begins as a response to external demand in relation to it: the higher H-SEA (or external environment) forms the demand for the results of the CA element. External demand is presented in Fig. 5.4 as an initial event "Received a message about the required performance results" event with the number 0.

The first phase of the LC CA is the design phase, and the first stage (elementary operation number 1 in Fig. 5.4) is the fixation of demand and awareness of the need.

The procedure of fixing the demand is described in detail in the previous section, so here we will model it as an elementary operation, without further detailing and structuring. This operation is presented as process, and not as decision-making because if the potential subject has made a decision not to carry out the activity, then the realization of the LC does not arise.

Therefore, in the model it makes sense to reflect only cases of positive decisions of the subject about participation in the activity.

Having realized the need, the subject defines the goal, structures it in the form of tasks (stage 2 in Table 5.1). Goals are formulated in terms of the expected characteristics of the results of the elements of the CA (for a detailed discussion of the concept of the result of the CA, see Sect. 6.1).

With a priori known demand and proven technology—regular activity—stages 2 through 6 are degenerate, they boil down to reading out the description of the logical and cause-effect model from the information store (see operation 2-a in Fig. 5.4). In other cases—the full execution of stages 2–6 on goal-setting and the creation of technology stages (logical and causal models, as well as resource pools). The goal-setting and the creation of technology for new elements of activity are of independent importance and, therefore, are separately presented in Sect. 5.3.

The next stage (7) forms a calendar-network graph. The consistency of key needs and resource pools is checked, optimization (stage 8) of resource use dynamics is carried out (taking into account the possibility of using these resources in parallel for the execution of other CAs). As a result, an optimal calendar-network plan and a schedule of resource utilization are obtained.

At the same time, the hierarchical structure of the A-SEA (downstream L-SEAs and L-Ops) is formed: the resource request and the designation of the subjects of the lower-level SEA are executed (step 9), as well as the request and assignments of the resources necessary to perform the elementary operations (step 10). The request, receipt and assignment of resources is carried out by the procedure described in

Table 5.1 Phases, stages and steps of life cycle of complex activity element (A-SEA) and their content

Phase	Stage	No.	Step	Content
Design	I. Fixing demand and understanding needs	1	Fixing demand and understanding needs	A superior U-SEA or environment forms the demand for the results of CA element. The subject (actor) fixes the demand, understands the needs and decides to perform activity
	II. Setting goals, structuring goals and tasks	2	Creating logical model	The need is structured and checked whether it is known or not (in the former case, the activity is regular)
				If CA is regular, this step comes to extracting information about the logical model from an information store
				Otherwise, the structure of goals is formed
				The goals are formulated in terms of the expected characteristics of the results of CA elements; see Sect. 6.1 for a detailed discussion of the result of CA
				Consistency is checked/the structure of goals is modified
				With each goal of A-SEA the role of the subject and technology is associated (this has been done for the result earlier); in other words, the characteristics of the subjects and technologies are specified.
				The result of this step is a logical model, i.e., the structure of A-SEA in the form of a set of subordinate SEAs (L-SEAs) and elementary operations (L-Op)
	III. Selecting and developing technology	3	Checking the readiness of technology and the sufficiency of resources	The presence of already known components of the A-SEA's technology is checked: the causal model of A-SEA, the technologies of all L-SEAs and the technologies of all L-Ops
				The logical consistency of A-SEA and resource pools is checked: the availability and sufficiency of resources for assigning the subjects of U-SEAs and supporting the technologies of L-Ops, taking into account the use of these resources in parallel when implementing other SEAs
				The result of this step is confirmation of the readiness of the technology, confirmation of the availability of necessary resources and transition to step 7, or the implementation of steps 4, 5 or 6, respectively

(continued)

Table 5.1 (continued)

Phase	Stage	No.	Step	Content
		4	Creating cause-effect model	The causal relationships between the goals/results of subordinate elements (L-SEAs and L-Op) are determined and described Possible events of uncertainty and the response rules for them are described (SEAs to be performed, or escalation to a higher level) The result of this step is the cause-effect model of A-SEA
		5	Creating technology of lower-level elements	For an elementary operation, due to its specificity and absence of internal structure, the process of designing and describing technology elements is specific and therefore has no general description For all subordinate L-SEAs without ready-made technologies, steps 1–6 of their life cycles are implemented recursively The result of this step is the technologies of subordinate elementary operations (L-Ops) and the technologies of subordinate L-SEAs
		6	Forming/modernizing resources	In the absence of necessary resources, goals responsible for their generation are set; SEAs ensuring the creation or modernization of resource pools are implemented The result of this step is resource pools required
		7	Calendar-network scheduling and resource planning	A calendar-network schedule is being formed. The consistency of key deadlines of needs is checked. The temporal consistency of the calendar-network schedule and resource pool is checked, taking into account the use of resources by other elements of CA In case of inconsistency, a return to steps 2–4 is carried out or the impossibility to meet the deadlines is escalated to the subject of a upper SEA. The result of this step is a calendar-network schedule for the use of resources
		8	Performing optimization	The dynamics of resources use is optimized, taking into account the possibility of using these resources for other CAs implemented in parallel The result of this step is an optimal calendar-network schedule for the use of resources

(continued)

Table 5.1 (continued)

Phase	Stage	No.	Step	Content
		9	Assigning actors and defining responsibilities	The responsibility matrix is fixed, which describes a correspondence between the subjects of SEAs and personnel. In fact, the assignment of subjects means the formation of demand for the results of lower SEAs and, hence, the recursive implementation of the life cycle of L-SEAs: all steps of the Design phase are carried out The result of this step is the responsibility matrix, which together with the structure of A-SEA determines its organizational structure
		10	Allocating resources	In accordance with the technologies of elementary operations, the resources required for the implementation of technologies are request and allocated The result of this step is the resource allocation matrix of elementary operations
Implementation	IV. Performing actions and obtaining results	11	Performing actions and obtaining results	In accordance with the causal model, the preconditions for the start of actions of elementary operations (L-Ops) and L-SEAs are repeatedly and constantly checked and they are launched The elementary operations (L-Ops) are performed The execution of subordinate L-SEAs is started The result of this stage is the execution of actions by A-SEA and also the result of its activity
Reflection	V. Assessing results and reflecting	12	Assessing results and reflecting	Comparison of the characteristics of the result with the required ones Comparison of the volumes of resources with the given ones Design of the requirements to the corrections of goals, technology, etc.

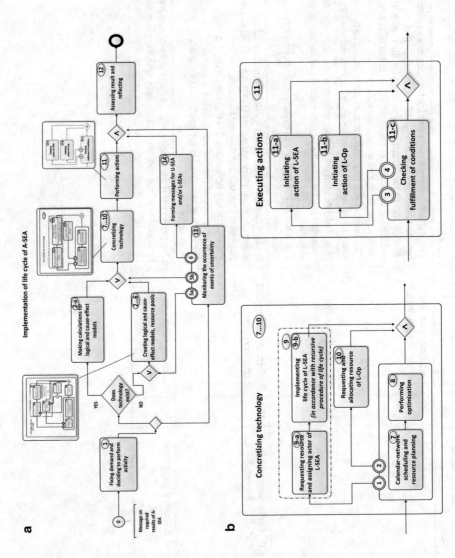

Fig. 5.4 a BPMN-representation of the process of A-SEA life cycle execution, b Elements of BPMN-representation of the process of A-SEA life cycle execution

Sect. 5.1 and is modeled by the elements 9-a and 10. The hierarchical structure is modeled with the help of the mechanism of undefined events and reactions to them: for each of the child elements of the L-SEAs and L-Ops event (shown in Fig. 5.4 as event-circles with numbers 1 and 2, respectively), and each of the events triggers the reaction in the form of elements 9-a and 9-b or 10. The goal of the subjects of the L-SEAs actually means the generation of the corresponding c millet, awareness of its future subject and execution of the life cycle design phase of the L-SEA. That is, at this point a recurrent reference is made to the process of performing the LC in relation to the lower SEAs (9-b in Fig. 5.4).

This completes the design phase, as a result, the whole hierarchy of CA elements is established, cause-effect relationships between the elements of the CA are established, the elements themselves are associated with specific calendar terms, all subjects and resources are assigned.

At the execution phase of the SEA, the *actions* are directly performed according the *technology*. Actions in the case of the SEA are multiple (they occur in each L-SEA and each L-Ops), and the order and conditions for their fulfillment are determined by the cause-effect structure reflecting the system-wide SEA technology. Thus, the *execution phase* (stage IV, stage 11) consists in *implementing the causal structure* over time. This implies "*Performing actions*" for each element of the causal structure in the form of initiating the execution of SEAs and operations according to preconditions. When checking compliance with preconditions, not only the execution of all the preceding elements is taken into account (see the explanations to Fig. 3.9, Sect. 3.4), but also the corresponding rules (conjunctive-disjunctive forms) associated with preconditions, completion of other elements of activity and/or conformity of factors external environment to specified values. The set of factors, methods of their control and other features of the stage are specific and do not represent an interest in this case. It is important that in some cases the actions can be performed, but in others this is impossible. In case of non-observance of preconditions, the subject can wait for some time (in a particular case, zero) and, if the conditions continue to be disregarded, interrupt the LC operations, regarding the situation as the occurrence of an event of uncertainty.

The immediate start of actions is carried out, as in the case of creating a structure through the mechanism of responding to uncertainty. This mechanism includes:

- event 3 and operation 11-a, which initiates the execution of lower-level L-SEAs (their main chains, starting from vertex 0),
- event 4 and operation 11-b, initiating the execution of lower-level operations of L-Ops.

The event specification rules (3 and 4) in this case will repeat the rules associated with the vertices of the cause-and-effect structure of the A-SEA, and the response rules are to determine which of the subsequent SEAs and elementary operations should be performed. In the example shown in Fig. 3.9 of Sect. 3.4, the event specification rule (type 3) for vertex f is defined as $(SEA_1 \wedge Op_1) \vee SEA2$, the response rule is the initiation of operation 11-a, which initiates execution of SEAj (Fig. 3.9) by setting its primary start vertex.

Upon completion of the action, the subject analyzes the result and performs a reflection (stage V, step 12), this stage is also specific. Actually, *reflection* is one of the forms of knowledge management (Sect. 2.6), since the results of the analysis can be reflected in the form of changes and additions to the A-SEA information model.

Consider the chain of uncertainty response shown at the bottom of Fig. 5.4, steps 13 and 14.

Specific manifestations of uncertainty, forms of events, ways of fixing them are specific.

The common thing is that to respond to uncertainty it is necessary:

i.   Detect all *events of manifestation of uncertainty*—the occurrence of external events, actual deviations from the prescribed technology (procedures for extracting technological information from the repository, allocating resources, reflecting), and so on.
ii. To react to events reporting them to the subject of a superior SEA to escalate the arising problems or initiate the relevant elements of the activity.
iii. In the special case, suspend the execution of all or part of the A-SEA actions and subordinate SEAs. The stoppage of A-SEA actions is performed by the subject directly. To stop subordinate SEAs, events that are indicators of the manifestation of uncertainty are generated, for example, in the form of corresponding messages, to which L-SEA subjects respond.
iv. Organize the execution of steps i-iii throughout all stages.

We formalize steps i-iv by blocks 13–14.

The reaction to the uncertainty can be carried out in various variants, the choice of which in the general case is determined by some a priori prescribed rules that are an integral part of the technology.

Checking the occurrence of events (block 13) is done by comparing the current conditions with a priori specified *rules of the event specification*, which describe what to regard as events of uncertainty and to what classes they are related. The option is selected in accordance with the *reaction rule*. The reaction rule puts in correspondence to the event of each class a certain "reaction":

- information escalated to the superior SEAs or transmitted to the downstream SEAs (block 14, following event 6),
- initiating the execution of lower-level L-SEAs or L-Ops (blocks 11-a or 11-b);
- stopping actions and returning to the revision of goals and changing the technology or re-forming the calendar-network and resource plans and assigning resources (consequence of events 5-a or 5-b).

The process model given in this section sets the template or basis for modeling the life cycle of an SEA.

So, the process model of a specific element of activity includes:

- system-wide basis—the cause-effect model of the LC in BPMN-notation, see Fig. 5.4 and Table 5.1;
- descriptions of system-wide SEAs (3… 6, 9-b) and system-wide elementary operations (9-a, 10, 11-a, 11-b, 11-in);
- descriptions of specific elementary operations 1, 2-a, 7, 8, 12;
- rules of event specification, and reaction rules.

The fractal properties of SEAs in the formalism of the process model are manifested in the recursive reversal of the process of the SEA's execution to itself, reflected by the element 9-b Fig. 5.4, which is also illustrated by Fig. 5.5.

It is worth noting that in the process model of the system-wide it is the integrating structure itself, the BPMN notation, and all the interfaces between the elements (inputs and outputs) and the internal mechanisms for implementing the greater part of the elements, that is, they are identical for the LCs of any SEAs.

Specific, that is, different for different SEAs, are only the mechanisms for performing operations 1, 2-a, 7, 8, 12 and the rules for responding to uncertainty. Descriptions of the specifics of these operations can be performed both in the natural language form and with the use of formal tools. The remaining blocks and operations do not require specification, since they are of a system-wide nature.

Let us consider how the proposed formalism can be applied to a concrete example about NPPs, started in Sects. 3.3 and 3.4, and supplement it with models for the execution of the SEA CA.

Performance of the activity "Carrying out works on the current maintenance of a specific unit of equipment of the turbine hall of the NPP" is reflected in the model of the SEA LC as follows.

Stage 1 operation 1 ("Decision-making to carry out activity"). Operational management of the NPP is assigned to the employee responsible for performing the work, and specific deadlines are determined based on a priori the drawn up schedule. At most NPPs, the schedule of such work is determined for several years ahead, specific dates are specified on an annual or quarterly basis. The appointed employee (A-employee) agrees with the assignment, performing block 1 (if there is no extraordinary events—uncertainty events).

Stage 2. The activity is repeatedly repeated and regulated, that is, regular. Therefore, Steps 2–6 are implemented in the form of Operation 2-a "Read Logical and Cause-Effect Model" and consist in extracting regulations and instructions for performing the work. The composition of the work is determined by a typical schedule and is also usually indicated in the order, the composition of the goals of the lower level, SEAs G_1–G_5 (Fig. 3.14) is thus determined.

Steps 7–10. The A-employee, on the basis of the deadlines set by the NPP management, forms the work schedule (operation 7) and optimizes ("leveling") the resources (operation 8). After this, the A-employee notifies all employees who must manage

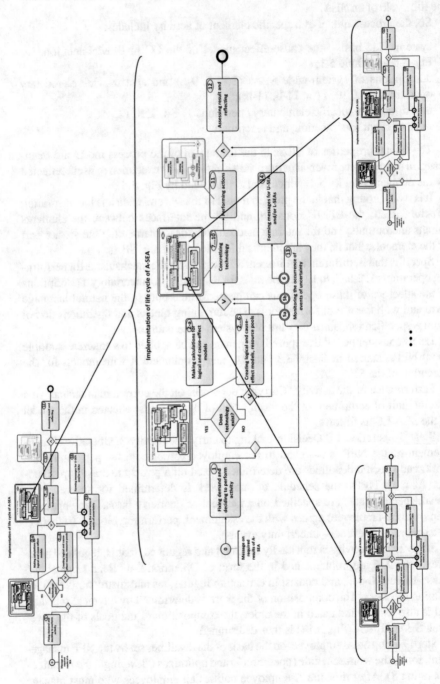

Fig. 5.5 The fractal properties of SEAs in the formalism of the process model

individual jobs, about the planned work. Thus, he realizes operations 9-a and 9-b with respect to the G_1–G_5 SEAs. The expression 10 for this SEA is not relevant.

Stage 11, operations 11-a, 11-b, 11-in (execution of the technological phase). In the shift and the time when the work should start on the withdrawal of the unit from the operating mode (the GD of G1 is the first of the "main chain"), the A-employee instructs the operators to start the output of the unit from the operating mode (precondition of the vertex 0 of the "main chain" Fig. 3.14). After completion of the output of the unit from the operating mode (precondition for the start of the execution of the SEA G2 Fig. 3.14) disassembly (SEAs G2) and diagnostics (SEAs G3) are performed. Upon completion of the work on the diagnosis and replacement of parts (preconditions for the start of the SEA G_5), assembly and testing under load. All the preconditions used above are attributes of the causal structure.

Stage 12 (Evaluation and Reflection). Upon completion of the work, relevant acts are issued, an analysis is made, the results of which are documented and stored in the archive.

Operations 13–14 ("The chain of reaction to uncertainty"). If, as a result of the diagnosis, the need to replace parts is clarified, then the corresponding work is performed. The indefinite event arose "inside the regular G_3 SEA", it was escalated by the SEA subject G_3 to the subject of the superior SEA—A-employee. The A-employee classified this information as an event of type u1, and, in accordance with the event specification rule and the response rule (example rules defined in Tables 5.2 and 5.3) initiated operation 11-b with the identifier of the initial vertex u_1.

During the execution of the SEA, other events of uncertainty may arise. For example, unforeseen delays in performance of work ("inside the G_1–G_5") or the inability to form a structure, due to illness of one of the employees (in operation 9-a received a message of refusal to carry out activity). In this example (Fig. 3.14 and Fig. 5.4), all these events entail escalating the problems to the higher level. However, the mechanism of the cause-and-effect structure allows you to specify the rules for "restarting" individual SEAs or chains of SEAs in the event of uncertainty events.

Table 5.2 Specification rules for events: current maintenance works for a specific unit of equipment of the turbine hall of the NPP

Specification	Type of event
Inadmissible deterioration of part N	Type u_1, argument N
Delay/stop of works G_1–G_5	Type 2; argument G_x
Rejection during formation of G_1–G_5	Type 3; argument G_x
Other deviations from plan	Type 4

Table 5.3 Response rules for events: current maintenance works for a specific unit of equipment of the turbine hall of the NPP

Type of event	Response
Type u_1, argument N	Initiate vertex u_1; argument N
Type 2; argument G_x	Escalate; arguments: type 2; G_x
Type 3; argument G_x	Escalate; arguments: type 3; G_x
Type 4	Escalate; arguments: type 4

Thus, the process model of the activity "Carrying out works on the current maintenance of a particular unit of the turbine room of the nuclear power plant" includes a typical BPMN model. Figure 5.4 together with Table 5.1, supplemented by specific text descriptions of operations 1, 2-a, 7, 8, 12 and the rules summarized in Tables 5.2 and 5.3.

Similarly, the execution models of the remaining illustrative examples discussed above can be described.

5.3   Model of Goal Setting and Creating Technologies for a New Component of Activity

One of the typical elements of the CA is the goal-setting and the creation of technologies. This element is the basis of development and provides the generation of new activity, so it should be considered in more detail.

5.3.1   System-Wide Properties of the Process of Goal-Setting and the Creation of New Technologies

System-wide analysis of the concepts of technology and related entities allows us to formulate the following four theses, characterizing the process of goal-setting and the creation of technologies.

Thesis 1 The process of goal-setting and creating a new technology is an activity, therefore all models and approaches developed above are applicable to it. This process can be represented as a set of SEAs and elementary operations, specified using models of Chaps. 3–5.

The choice of one of the SEA formalisms or an elementary operation for the representation of an activity element implements the principle of generalization and abstraction (Sect. 1.2) and is carried out on the basis of the following considerations. If some element of the CA possesses an internal structure of a system-wide character, then it is modeled by the SEA, otherwise by an elementary operation. This approach ensures the creation of a model that reflects the system-wide structure and system-wide characteristics, and also allows for further detail when it is necessary to build a model for the process of creating a specific technology on its basis.

Thesis 2 *Technology* was defined in the introduction as a system of conditions, criteria, forms, methods and means of successive achievement of the stated goal. Based on the analysis of the concept of technology performed in [97], this definition can be detailed.

The above definition underscores the "purposefulness" of CA technology, its orientation toward achieving the goal. This definition defines not just a lot of elements of technology, but a certain ordered order of their totality, that is, it represents technology as a system. The definition also specifies the following elements of technology:

i.   specific conditions in which the CA is implemented;
ii. forms of organization of CA;
iii. methods as a concept, generalizing operations and techniques;
iv. means of implementing CA;
v.   criteria for achieving the goal.

Thesis 3 All elements of technology are informational ("knowledgeable") or material. The conditions (element i) are a description of the circumstances under which the CA rules are implemented. That is, the condition is an information object, but the object of this description can, naturally, be real objects. A similar statement is valid for forms (ii), methods (iii), and criteria (v). The means for implementing CAs (iv) are, first of all, resources (see Sect. 2.9). Thus, elements i-iii and v are only informational, and iv are both informational and real. Consequently, the goal setting and the creation of new CA technologies can be represented by two processes, one of which (a) is aimed at operating information, and the second (b)—material objects:

(a) design and specification of the goal, conditions, forms, methods, means (including resources) and criteria,
(b) a certain organization of material resources (information resources are organized during the design process).

Since information on activity is contained in the information model, it can be argued that the goal setting and the creation of technology is an activity, the subject of which is always a "new" (for example, more detailed) information model, which establishes a correspondence between i and iii and resources, and in particular cases— also the resources themselves.

Thesis 4 The necessary element of the goal setting and technology creation process is the forecast of how the implementation of technology will ensure achievement of the objectives of the activity under the influence of possible uncertainty. This is due to the fundamental feature of creating a new technology, consisting in the separation in time of the periods of its creation and use in the course of the realization of activity. The creation of technology corresponds to the initial stages of its life cycle. *Forecasting* can be very complex and can be considered as an independent complex activity. This process is specific, but the influence of uncertainty for forecasting goals can and should be structured according to the two system-wide grounds discussed above. First, they are *primary groups of sources of uncertainty* (see Sect. 2.7):

• uncertainty of the external environment—external demand and external conditions, requirements and norms;
• uncertainty of technology and subject matter—means, methods and factors;

- uncertainty of the subject—awareness of the external need, goal-setting, execution of actions, evaluation of the result, and decision-making, whether to act as a CA subject.

Secondly, these are structuring *binary features* (Sect. 4.4):

- Will the result have planned properties/functions?
- Will the resulting properties/functions provide the desired effect in interaction with the external environment?
- Will the properties of the result correspond to the demands of the demand at the future moment of presenting the result to the consumer?
- Will the external environment (at the moment of presenting the result to the consumer) correspond to the current forecast of its state?

On the one hand, technology is a set of interrelated elements of i–v, and on the other, any SEAs can be described by the models of structural, cause-effect and process models used in this paper. The fractality of the elements of the CA requires (for the completeness of the representation) also modeling the elements of the CA that are inferior to the logical structure. Such a set of models describe the conditions, forms, methods, means and criteria for achieving the goal, that is, elements of technology. Therefore, at the system-wide level of generalization, the following statement can be formulated.

Composition of the system-wide technology of the complex activity element.
The technology of any CA element is described by the structural, cause-effect and process models together with models of subordinate SEAs and elementary operations.

5.3.2   Model of Goal Setting and Creation of New SEA Technology

We now form the model of stages 2… 6 (Table 5.1) of the life cycle of the CA element, which ensure the goal setting and creation of the new CA element technology (let's call it A-SEA). The model will be implemented in the BPMN notation (Fig. 5.6), based on the modeling principles introduced in the previous Sect. 5.2.

The first operation (2 in Fig. 5.6 and step 2 in Table 5.1) is the formation of the logical structure of the sub-goals of the A-SEA.

Subgoals are formulated in terms of the characteristics of the expected results. Each of the subgoals requires for its achievement the realization of an associated elementary operation or an SEA. The elements of the CA of the lower level with respect to the A-SEA level are called L-Ops and L-SEAs.

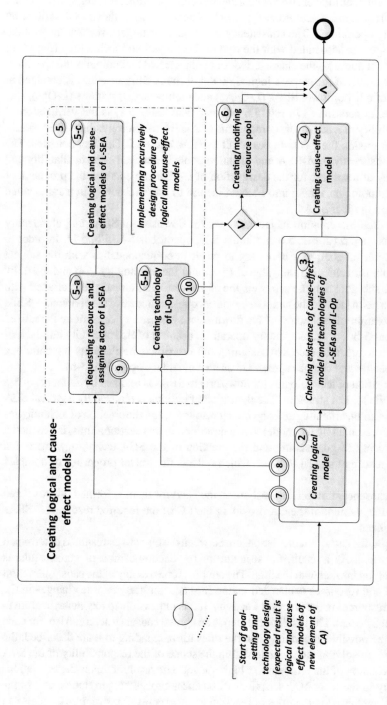

Fig. 5.6 Causal structure of «Goal setting and creation of the new CA» of C-SEA

The direct definition of A-SEA goal and its decomposition into subgoals (L-SEAs) are specific actions, therefore at this level of generalization they are described by elementary operation 2. The consistency is checked/the target structure is modified. Each sub-goal is associated with the role of the subject and technology (the result is already set above)—the characteristics of subjects and technologies are specified. Thus, as a result of the stage, a logical model of the A-SEA structure is obtained in the form of a list of lower SEAs (L-SEAs) and elementary operations (L-Ops).

The next operation (3 in Fig. 5.6 and step 3 in Table 5.1) is the verification of the availability of ready-made components of A-SEA technology: the A-SEA cause-and-effect model, the technologies of all L-SEAs and all L-Ops technologies. The logical consistency of A-SEA and resource pools is checked—the availability and sufficiency of resources for the designation of L-SEA subjects and the provision of L-OPs technologies, taking into account the use of these resources in parallel when implementing other SEAs.

In parallel, the creation of technologies for downstream SEAs and elementary operations, step 5 (in Fig. 5.6 and Table 5.1), operations 5-a, 5-b, 5-in. In order to achieve each subgoal, it is necessary to put it in correspondence with the subject responsible for achieving the subgoal (5-a), and the technology (5-b and 5-in), by which the sub-goal must be achieved, the object/result has already been aligned with the goal in point (1), and the target tree—an adequate structure of the elements—SEA.

The creation of technologies for downstream elementary operations is realized by element 5-b. For an elementary operation, because of its specificity and lack of internal structure, the process of designing and describing the elements of technology is also specific and can be represented as an elementary operation (5-b).

The creation of technologies for downstream SEAs is realized by elements of 5-a and 5-in the logical structure. The definition of entities is realized by a typical SEA (5-a) requesting, obtaining and organizing resources, and the creation of technologies (5-c) is represented by the model that is described in this section. Thus, in the model of the process of goal-setting and the creation of the SEA technology, there is a multiple recurrent appeal to itself, which reflects the fractal properties of complex activity.

In the absence of necessary resources, the need for them is formed (events 9 and 10), and the element of the CA providing the LC of the required resources—SEAs 6 is organized.

In step 4, the cause-effect relationship between the objectives/results is determined and described. As a result, the structure of the decomposition of goals/results is formed taking into account the time. The temporal consistency of the causal-temporal structure and resources is checked taking into account the possible changes in the states of resources in the process of activity. If it is impossible to reconcile, a return to operation 2 occurs. The result of the step is an agreed causal structure. Also, for each L-SEA, the possible uncertainty and the rules for responding to it are described: the procedure for solving the problem within the scope of the responsibility of the SEA or the escalation of the problem to a higher level. The result of this is the regulations of activity in the form of descriptions of business processes, job duties, etc. In the course of synthesizing the rules of reaction to uncertainty, the structure of goals can be modernized, which causes a return to subgoal 2. The operation of designing a causal structure is specific.

System-wide event specification rules and reaction rules in this model provide the creation of downstream technology through events 7 and 8, and the creation of new resource pools, events 9 and 10.

As an example of the execution of the model, we will consider the goal setting and the creation of a new technology for the activity "The introduction of polymer composite materials (PKM) into the design of the MS-21 medium-haul airframe" [124]. Composite materials are widely used in modern aircraft building, in particular, in the design of the airframe of the Boeing 787 aircraft, the share of aggregates of them is more than 50% by weight, Airbus A380–30%, Airbus A350–50%.

The subject of the activity (G_t's SEAs in Fig. 5.7) for the execution of the PCM is the engineering center of Irkut Corporation, which executes the design of the product (MS-21 aircraft), which organizes and manages the cooperative to create the "expanded enterprise" product (Sect. 2.4). The introduction of PCM affects all stages of the life cycle of the future product and most of the major participants in the cooperation, so the implementation problem was structured (by the engineering center specialists) for a number of tasks [124], each of which was decided by the respective enterprises involved in the cooperation. Thus, the first element of the activity on the creation of technology—the design of the logical structure (2 in Fig. 5.6) is completed by the definition of goals for the elements of the activity of the subordinate level:

1. choice of brands of materials used and their suppliers;
2. definition of the requirements for the design of aircraft units;
3. development of the scheme of production cooperation;
4. selection and development of production technologies;

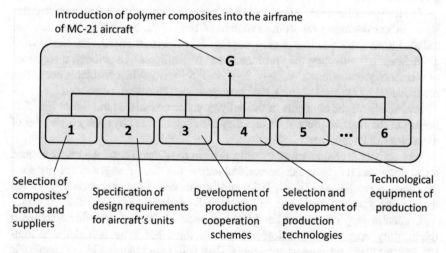

Fig. 5.7 The logical structure of SEA "Implementation of PKM in the design of the airplane MS-21MS-21"

5. technological equipment of production;
6. and so on.

The logical structure of the SEA "Implementation of PKM in the design of the airplane MS-21" is presented in Fig. 5.7.

In the course of creating technologies that are lower in the logical structure of SEAs and elementary operations (5-a, 5-b, 5-c in Fig. 5.6), the subjects of these elements of activity are first determined: the first three elements (1–3 in Fig. 5.7) are performed by specialists engineering center, and the other two (4 and 5 in Fig. 5.7) suppliers of aggregates (CJSC "Aerocompozit" CJSC "Aviastar SP", JSC "ONPP Technology").

The development of technology elements (5-b in Fig. 5.6) "Selection of grades of applied materials and their suppliers", "Determination of requirements for the design of aircraft units" and "Development of production cooperation schemes" is carried out by specialists of the engineering center in the form of plans and regulations of the relevant work.

The organization of resources for these works (element 6 in Fig. 5.6) is reduced to additional training of specialists of the engineering center, as well as providing them with appropriate tools and information resources.

The request and receipt of resources that provide subjects of lower-level CSAs (element 5-a in Fig. 5.6) is realized in the form of coordinating the participation of enterprises—suppliers of units in cooperation and securing cooperation ties with appropriate contractual and technological documents.

Technologies of downstream elements ("Selection and development of production technologies" and "Technological equipment of production", 4 and 5 in Fig. 5.7) are developed by the subjects of these elements (suppliers of aggregates) in the form of a recurrent execution of this model for creating new technologies of activity (element 5-in on Fig. 5.6).

The design of the cause-effect model (4 in Fig. 5.6) is of interest at more detailed levels of consideration (as in the example of Sect. 5.2), but at this level they are reduced to an intuitive-clear result—see Fig. 5.8. The causal model has the form of a "main chain" reflecting the "problem-free" realization of an activity: a sequence of elementary operations $1 \rightarrow 2 \rightarrow 3 \rightarrow 4 \rightarrow 5 \rightarrow$ …. When problems occur that are modeled by events 1, 2, or 3, the corresponding returns are performed.

Practically all the elements of the activity under consideration (1–5 in Fig. 5.7) belong to the creative (Sect. 4.4), and they are characterized by a very high level of technological uncertainty.

Any activity is characterized by reflection: in the general case, the newly created technology and its result can be verified and/or falsified in any way (at the epistemological level—see [71, 74, 108]) and/or checked at the execution level of the technology. Such a "check" can be conditionally called a *technology test*.

CA technology on the one hand has a significant share of specific components (elementary operation technologies), and on the other hand it includes a logical, cause-effect and process models of CAs and, consequently, has system-wide components.

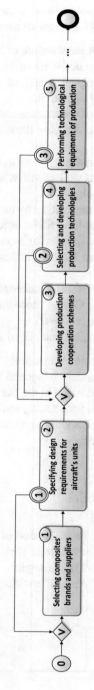

Fig. 5.8 The causal structure of SEA "Implementation of PKM in the design of the airplane MS-21MS-21"

Therefore, testing the technology, including the completeness of its description, also has *system-wide features*; we formulate them in the form of a list of verifiable conditions that must be met when creating a new technology:

I.   Completeness and consistency of the structure of the "technological goals" of the logical structure, the structure of the sub-objectives of the Ts-SEA.

II. Availability for each "technological goal" of the specified subject and technology—SEA or elementary operation.

III. The presence of a specification of the characteristics of the object of the CA of each of the subgoals, allowing verifying the achievement of sub-goals and effectiveness.

IV. Consistency (mutual logical consistency) of actors and technologies associated with the structure of sub-goals, as well as with resource pools required for the organization of actors and technology.

V.   Presence of a specification of the mechanism for assigning resources to perform the functions of subjects of lower-level SEAs or to ensure the technologies of subordinate elementary operations, including ensuring consistency of individuals' goals and preferences when they are assigned by SEA subjects to the objectives of the C-SEA.

VI. The presence of specifications of events of uncertainty and the rules of reaction to it within the framework of the C-SEA (decision-making procedures or the escalation of the problem to a higher level).

VII. Consistency and coherence of logical and causal structures.

Inspections I–VII can be implemented collectively within the framework of a separate reflexive operation or be separately integrated into separate operations, for example, checks I and II can be implemented in the framework of operation 2.

One of the important elements of the goal-setting and technology creation activity is the organization of new resource pools and the realization of their life cycles, so let's move on to modeling these elements.

5.4   Resources Life-Cycle Model

Resources are an aspect essential for the realization of activity, without modeling, the study of complex activity will be incomplete. Therefore, this section is devoted to the modeling of system-wide factors in the organization of resources, life-cycle management of resources, a particular case of "auxiliary" activity.

The organization and execution of the life cycle of resources was defined in Sect. 2.9 as an activity:

- to establish and maintain the structure of the functionality of resources;
- on provision of resources for technology provision, formation of the subject and subject matter of activity.

At the same time, quantitative and qualitative characteristics of resources should be provided according to the needs of the activity for which the resources are intended. Resource pools are organized aggregates of resources that are in interrelations with "core activity" and with each other.

To build a system-wide model of *resource management*, it is advisable to present the latter in the form of a single SEA that implements the lifecycle of the resource pool (see Fig. 5.9). Typical LC (see Fig. 5.9) defines the stages of this CSA, and their system-wide properties are discussed below.

It makes sense to note that during all these stages, not only the *resource pool* is organized, created, tested, modernized, contained and disposed of as an organized collection of their instances, but also the procedures for operating with resources. In fact, a *specific technology for organizing resources* is being created and is being implemented.

Analyzing the distribution of effects and costs by stages of the life cycle (similarly to what was done in the case of the LC of the CA, see Table 1.2), one can say that the resource effect only occurs at the stage of "providing resources," while costs must be borne during all stages. At the same time, the costs ratios at different stages can vary greatly depending on the specific resources.

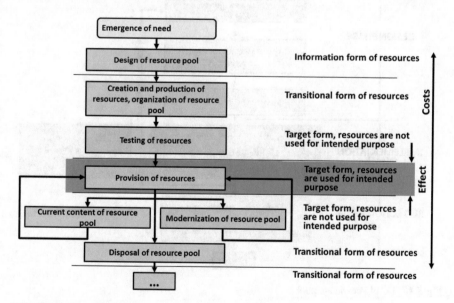

Fig. 5.9 LC of resource pool

An important special case of resources is *human resources*. Section 2.4 noted the features of people's behavior as elements of CA subjects and the need to take into account the influence of feelings, emotions and other "transcendental" factors on how people conduct activity. In this sense, it is advisable to use a widespread in the management hike, when the activity of an employee in the firm is characterized by loyalty and productivity, and all work related to human resources (HR function) is aimed at ensuring long-term and productive functioning of employees in the firm. All the components of the HR function correspond to those defined in Fig. 5.9 stages and phases of the resource center. For example, at the stage of creating a pool of resources, personnel planning, hiring, training of employees is carried out, and at the stages of maintenance and modernization—training and development, evaluation and promotion, employee motivation and other components of the HR function.

Another particular case of resources is knowledge (see Sect. 2.6) and technology (technological knowledge as "operational knowledge", see Fig. 5.10).

The life cycle of the knowledge itself, as an object, is simple and includes *three phases*:

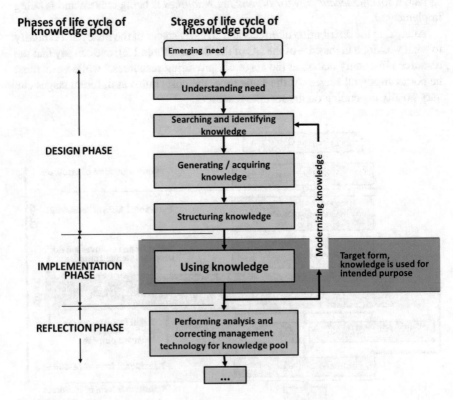

Fig. 5.10 LC of knowledge pool

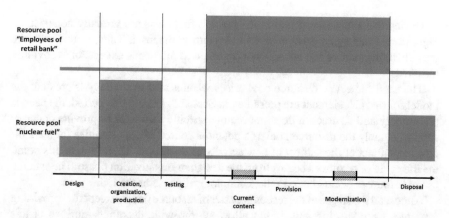

Fig. 5.11 Examples of costs allocation on LC stages

1. creation/formation, analog of the organization and design phase;
2. productive use (with a possible return to the first stage), an analog of the technological phase or execution;
3. an infinite existence in the form of historical data (with a possible return to the second stage).

In the case when knowledge is used as an element of technology, as "operational knowledge", the life cycle of the pool of such knowledge (see Fig. 5.10) is similar to the lifecycle of the resource pool shown in Fig. 5.9.

As two illustrative examples, when costs are distributed almost directly in the opposite way, it is possible to result in a resource pool of employees-client managers of a retail bank or a pool of nuclear fuel resources of a nuclear power plant. In Fig. 5.11 shows conditional costs distribution diagrams for the stages of the LC of these resource pools.

Functional duties of client managers do not require their special training, the organization of their work is quite simple, the profession is mass, so recruitment does not present problems, and there is no need for additional training. Therefore, designing a pool of resources in the form of writing rules for the actions of the client managers themselves and the employees who manage them is not laborious; creating a pool in the form of hiring the appropriate specialists also requires costs substantially less than the costs of maintaining employees. That is, in this case the overwhelming share of costs falls to the stage of "provision of resources".

In the case of nuclear fuel, the costs of designing a new type of fuel is relatively small. The major part of the costs is accounted for ore extraction, enrichment, fuel fabrication, manufacturing of fuel elements and assemblies, transportation and storage.

During the direct use of fuel in the reactor, fuel costs are virtually nonexistent. Significant costs again arise after its use—during disposal. Thus, in this case the overwhelming part of the costs is concentrated at all stages except for "providing resources".

The need for a new resource pool arises when a new technology is created; the models of this CA element are presented in Sect. 5.3. During this period, the need is preliminary and abstract: it does not require specific resources to provide concrete actions, namely the resource pool as a potential source of opportunities for technology, the subject or the subject of new activity in the future. Therefore, at this point, the life cycle of resources begins from the design or organization phase. The specific use of resources occurs later, in the technological phase of the LC.

Almost all the stages of the resource center of resource pools, except for "providing resources," are specific and do not allow system-wide decomposition, so at this level of generalization they are modeled not by detailed goals and corresponding elementary operations.

The logical, causal and process models of the SEA "Realization of the life cycle of resources" are presented below. As in all the previous sections of this chapter, the simulated SEAs is referred to as the A-SEA, and its downstream elements are referred to as L-SEAs and L-Ops.

I. Logical model of the A-SEA "Realization of the life cycle of resources" (Fig. 5.12). The target (logical) structure of this SEAs is formed from the following system-wide considerations.

The life cycle of the resource pool begins in response to the need for activity to form a new technology. At the first stage of the resource center, a resource needs analysis is performed, the specific requirements for the resource pool are formed on its basis, and the resource pool is designed together with the technology of resource

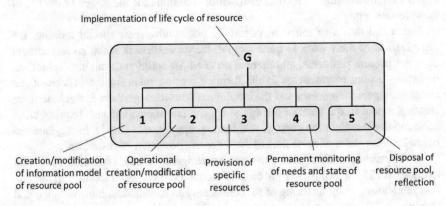

Fig. 5.12 The logical structure of A-SEA «Execution of resources LC»

organization. The goal of this stage is to form the first version of the information model of the resource pool that meets the needs (sub-goal 1 in Fig. 5.12).

After this and/or in parallel, the activity for the direct creation of the pool are carried out: recruitment, training and organization of personnel, when it comes to human resources; purchase, production, transportation, storage of raw materials, materials, tools or equipment; erection of buildings and structures; organization of information resources. We define the goal of this activity as the operational creation of a resource pool according to the information model (sub-goal 2 in Fig. 5.12).

The goal of the main stage of the resource center's resource center (subgoals 3 and 4 in Fig. 5.12) is to provide resources to meet specific needs arising from specific elements of the activity that use these resources. This goal allows and requires a system-wide decomposition, which is described later in this subsection.

In parallel with the provision of resources, activity is carried out to maintain and modernize the resource pool. From the system-wide point of view, the activity of both these types are reduced to changing the information model and the resource pool and the operational change of the resource pool itself with subsequent verification, that is, the goals of this activity are particular cases of subgoals 1–4 on Fig. 5.12. Content and modernization are reactions to the discrepancy between the current values of the characteristics of the pool of resources and needs (current or future). Therefore, the activity of both these types is modeled as the repeated formation/modification of the information model and the operational creation/change of the resource pool with subsequent verification of the achievement of the correspondence of the received characteristics of the pool to the required values. The resource center's resource pool ends with an element of the activity of disrupting the resource pool (sub-goal 5 of Fig. 5.12) and reflecting the evaluation of the results of the resource center's resource pool. In some cases this element is degenerate.

The sub-goal "provision of resources" is achieved by elements of activity 3 and 4 (Fig. 5.12), which have the following meaning. The goal of element 4 is to perform an operational control of the needs and status of the resource pool. The meaning of activity element 4 is to record the specific needs of other related activity during the whole LC stage and initiate the provision of resources, as well as to track the matching of the characteristics of the resource pool to the needs. The need for operational tracking is explained by the possibility of changing, firstly, the characteristics of resources under the influence of the external environment and own behavioral features over time, and secondly, by changing the need itself. Element 3 provides a response to specific needs, and the response to the disagreement between the characteristics of resources and needs is realized by reapplying elements 1–3.

II. Models of children below in the logical structure of CA elements. Because of the specificity of the sub-goals, the activity to achieve each of them within the framework of this model are not detailed and are represented by elementary operations 1–5.

III. The causal model of the resource pool LC (Fig. 5.13). The general logic of achievements of sub-objectives of the A-SEA is determined by the stages of the

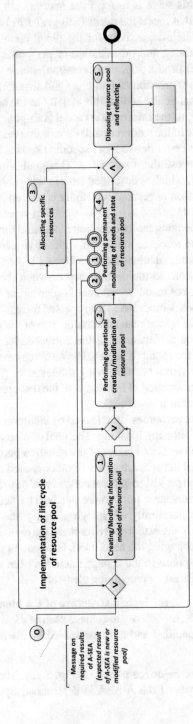

Fig. 5.13 The causal model of the resource pool LC

lifecycle of the resource pool; therefore the sequence of elementary operations 1 → 2 → 4 → 5, supplemented by the initial and terminal events, forms the main chain.

As in the case of creating a new technology, this element of activity is characterized by uncertainty, which entails not only negative, but also positive consequences. Constructive character is the event of occurrence of a specific resource requirement.

We simulate the fixation of the demand and the reaction to it by the chain < event 3 → operation 3>. If the discrepancy between the current values of the characteristics and the needs of the response is found, in the general case, the re-design of the resource pool is a modification of the information model (operation 1) followed by the operational implementation of changes in the resource pool (2). We model these reactions by the chain

<event 1 → operation 1 → operation 2>.

In case of unsuccessful operations 1-4, it may be necessary to perform them again— the chain

<event 1 → operation 1 → operation 2>

and

<event 2 → operation 2>.

Note that the activity for the organization of resources is characterized by the properties of the dispatching of passive replicative activity (Fig. 4.5 of Sect. 4.3).

IV. The A-SEA process model. The process model of the SEA "Realization of the life cycle of resources", as well as the model "Targeting and creating a new technology," is a model, describes not a specific activity, but a class of activity. For the same reason (see Sect. 5.3.2), it makes no sense to detail the descriptions of the specific elementary operations of the process model.

Similarly, system-wide event specification rules and reaction rules (lines 1 and 2 in Tables 5.4 and 5.5) in this model not only control the repetitions of the "main chain" (events 1, 2), but, above all, ensure the fulfillment of the main goal A resource pool is the provision of resources through event 3 and operation 3 (Fig. 5.13).

The main goal of the resource pool is implemented by the chain <event 3> operation 3>, rules 3 in Table 1.3 correspond to it. Tables 5.4 and 5.5. The rules declare that when a message is received about a particular resource requirement, operational activity for their provision are carried out (3).

Table 5.4 Specification rules for events: A-SEA "Implementation of life cycle of resources"

	Specification	Type of event
1	<specific conditions 1>	Type 1;
2	<specific conditions 2>	Type 2;
3	Receiving message about concrete need in resources	Type 3;
4	The number of repetitions or the total execution time exceed th thresholds	Type 4;
5	<other deviations from planned execution >	Type 5
...	...	...

Table 5.5 Response rules:
A-SEA "Implementation of
life cycle of resources"

	Type of event	Response
1	Type 1;	Initiate operation 1;
2	Type 2;	Initiate operation 2;
3	Type 3;	Initiate operation 3
4	Type 4;	Escalate; argument = type 4;
5	Type 5;	Escalate; argument = type 5;
...	...	...

The fourth line prevents unlimited repetition of the main chain in case of unsuccessful execution.

The fifth lines contain rules that determine the general case of a reaction to other significant deviations from the regular performance of an element of activity.

Thus, a model model for realizing the life cycle of resources in the form of an SEA is formed, the model includes the following system-wide objects:

- logical and cause-and-effect structures, Figs. 5.12 and 5.13, as well as the rules associated with intermediate vertices (in the text);
- the process model in BPMN notation, Fig. 5.4, event specification rules and reaction rules, Tables 2.4 and 2.5.

Specific objects integrated into the general structures of this model allow for detailed elaboration in the formation of the model of a specific element of the organization of resources.

Consider the application of the considered model for realizing the life cycle of resources for solving practical problems by the example of one of the engineering divisions of some company of an aircraft building company engaged in the development of chassis or fuel systems or air conditioning systems, etc.

From the point of view of the formalisms used, the functioning of such a unit during any period of time represents, on the one hand, the implementation of several elements of the "core" complex activity (the subjects of these SEAs are employees, possibly organized in working groups), on the other hand, resources (employees) that provide "core" activity. Let us analyze the second aspect of the unit's operation. The subject of "auxiliary" activity for the implementation of the data center of these resources is the head of the unit. Usually the operation of such a subdivision is managed and organized on the time horizon for one year with clarification and adjustments for quarters and months. For each of these intervals, a specific model of the activity element, formed on the basis of the model model (given in this section above) can be presented by detailing and specifying the specific Op_1–Op_6 operations according to the general method of generalization and abstraction (see Sect. 1.2).

Thus, the SEAs "Operation of the unit during a specific reporting year" (execution of the life cycle of resources during the year) is described by the models Figs. 5.12, 5.13 and Tables 5.4 and 5.5, and the elements of activity presented in these models as elementary operations (1–5), should be detailed and specified.

Such an activity begins with the SEA "Create/Change the Information Model of the Resource Pool" (Figs. 5.14 and 5.15).

In this case, the information model of the resource pool exists in the form of the unit's staffing schedule, job descriptions of employees, regulatory documents of the unit, technological documents describing the regulations, methods, methods of work of employees, and the tools used.

First of all, according to the plans of the Company or the higher-level subdivision, including taking into account the prospects for development, it is necessary to update the forecast (1–1, Figs. 5.14 and 5.15) of the resource needs in terms of competencies and the number of employees, on the basis of which technological and organizational documents, personnel plan, including plans for the recruitment and dismissal of employees, training and retraining, the development of new (adjusting existing) technology work.

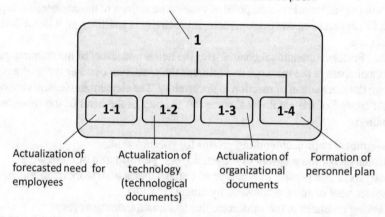

Fig. 5.14 The logical structure of SEA «Create/modify the information model of resource pool»

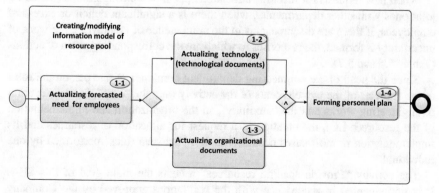

Fig. 5.15 The causal structure of SEA «Create/modify the information model of resource pool»

The actualization of technological documents (1–2, Figs. 5.14 and 5.15) is in fact a correction of the technology of the division's employees, and the actualization of organizational documents (the formation of a personnel plan) is a correction of the manager's technology. Therefore, all these types of activity are implemented in the form of SEAs based on the model model for the creation of new technologies, given in Sect. 5.3. At the same time, the development of technology elements affecting other parts of the company is carried out in the form of separate SEAs, the subjects of which are joint working groups. Such elements are, for example, specific design and technological work carried out in coordination with other engineering units, and training, hiring and firing of employees, carried out and/or organized by personnel units.

The element of the activity "Run operational creation/change of the resource pool" is a routine execution of the formed personnel plan throughout the year. Detailed modeling of this activity in the form of a SEA or a hierarchy of SEAs can be of interest only when developing and optimizing the business processes of personnel management-specific regulations for training, hiring, firing employees and other personnel operations. From the point of view of the activity of the engineering department, its organization and management, this activity is auxiliary, so it is not detailed in this case.

The "Perform operational control over the needs and status of the resource pool" activity element is the manager and initiates the execution of other types of activity through the mechanism of reaction to uncertainty. The element implements the operational control of the employees' status on the part of the head of the subdivision, including:

- assignment to project/working groups for specific work,
- an analysis of the progress of the employees' performance of these works,
- analysis of compliance of employees' competencies with work requirements,
- replacement of some employees by others,
- tracking absences in the workplace due to illness, training or leave,
- monitoring the availability of unassigned employees ("free resources") and so on.

When new requests for resource allocation appear in working groups (including joint ones with other departments), when there is a significant deficit or excess of employees, if there are discrepancies in the competence of employees, the work of uncertainty is formed, the responses to which are the corresponding chain of actions (Tables 5.6 and 5.7).

Since the head of the engineering department combines in one person the roles of the subjects of the two elements of the activity (and the "core", in the execution of engineering works and the "auxiliary", in the organization and implementation of the resources LC), the creation of a request for allocation of resources and its implementation in most cases occurs virtually: "between roles" performed by one individual.

The activity "Provide specific resources" reflects the main goal of this SEA. It is implemented in accordance with the regulations approved by the Company and consists in appointing employees to the appropriate working groups. This is an

Table 5.6 Specification rules for events: SEA "Implementation of life cycle of resources"

	Specification	Type of event
1	Absence of an employee in the workplace due to illness	Type 3;
2	Noncompliance of employee's competencies with work requirements	Type 2;
3	Voluntary termination of employment	Type 2
4	The number of repetitions or the total execution time exceed the thresholds	Type 4;
5	Receiving request for new employees to carry out works	Type 3;
...	...	...

Table 5.7 Response rules: SEA "Implementation of life cycle of resources"

	Type of event	Response
1	Type 1;	Initiate operation 1
3	Type 3;	Initiate operation 3
4	Type 4	Escalate; argument = type 4;
...	...	...

element of organizational activity to establish links between employees in order to form stakeholders.

The activity "To disband the resource pool and to implement the reflection" is relevant within the framework of this SEA only in terms of reflection—the evaluation of the department's results for the reporting period. In practice, such an assessment precedes the planning for the next period, therefore it makes sense to consider it within the framework of the SEAs to update the technological and organizational documents that are components of the activity to create and change the information model of the resource pool.

Thus, all elements of the SEA "Functioning of the unit" are specified and detailed at the level of the models under consideration, while the element "Create/change the information model of the resource pool" was expedient to present in the form of an SEA combining several other SEA based on a model model for the creation of new technologies. The remaining elements do not require structuring, therefore they are left in the form of elementary operations, and the details of their execution are described in text form.

All elements of the SEA "Functioning of the unit" are integrated into a single system by the logical, cause-effect and process models described in this section above and presented in Figs. 5.12, 5.15 and in Tables 5.6, 5.7.

The SEA model "The functioning of the unit during a specific reporting year" reflects the characteristics inherent to any elements of the organization of and management of any human resource pools. Such features are the following:

- The activity under consideration includes the development or updating:

    (a) technology of core activity that is carried out by human resources;
    (b) technologies for organizing and managing the pool of resources;
    (c) the development plan for the resource pool.

- The key elements of this activity are "Execution of operational control of needs and conditions of the resource pool" and "Provision of specific resources".
- "Execution of control" includes, first of all, verification of the set of conditions describing the resource pool, their activity and requests for resources.
- "Provision of specific resources" is a standard procedure, so it is advisable to regularize it, fixing the technology in the form of regulations.
- Usually, the resource pool is used long-term for many successive periods, so the final element for disbanding the pool and reflecting is realized only in the part of the reflection. Moreover, the reflection of the activity of the completed period usually serves as a basis for updating the technology of activity in the next period; therefore it makes sense to implement it during the actualization of the technology.
- Functioning of the resource pool is also usually considered in the context of periods of different lengths—year, quarter, month. Therefore, each of these periods has its own elements of activity that naturally intersect and, therefore, should be considered together.

For effective realization of activity, these features should be, firstly, realized by the head of the subdivision in question, and secondly, to some extent fixed in the relevant regulatory and regulatory documents that are in fact a form of representation of the information model.

In practice, the activity of the units is regulated by a combination of such documents as regulations on the division, job descriptions of the manager and employees, regulations for planning and reporting units within the organization, as well as regulations for interaction between units and others. Therefore, it makes sense to reflect the model of the process of realizing the life cycle of resources and its features as accurately and fully as possible. This will minimize the involvement of management (escalation of problems) and thus reduce the costs of performing activity—the implementation of the resource center.

In fact, this description of the SEA "Operation of the unit during a specific reporting year" represents the target ("ideal") model for implementing the life cycle of the unit (resource pool) during the reporting period. Based on this description, you can create a list of control questions (see Table 5.8) that allow you to formalize, structure and evaluate the differences in the level of organization and management of a particular unit from the target level.

Table 5.8 Control questions to identify differences between the current level of organization and management of a specific department and the target level

No.	Question
Information model of department's life cycle	
1	Is there a description[a] of the technology of main activity: – the categories of CA implemented by the employees of the department? – planned/promising CA? – the competencies of employees needed to fulfill each of the categories of CA? – other resources (besides employees) needed to implement each of the categories of CA?
2	Is there a description of resource demands, including: – a specific subject of activity forming a demand? (The activity has been described in item I.1.a.) – historical data on resource demands for previous periods? – estimated resource demands for future periods? In what forms and for what time periods?
3	Is there a description of the processes of replenishing and disposing the resource pool: hiring? training? firing?
4	Are there any descriptions of resource allocation processes, e.g., for assigning employees to working groups, for assigning jobs to employees, etc.?
5	Are there descriptions of the actualization processes of the information model elements, items 1–4?
6	Do the processes in practice correspond to the descriptions of items 1–5?
7	How often is the description of items 1–5 actualized?
Practical implementation of department's life cycle	
8	How is operational information fixed? How can it be acquired? How relevant is it? Operational information covers the following aspects: – the current composition of the department (resource pool) – the employees assigned to working groups for particular jobs; – planned assignments of employees to working groups and their allocation for jobs (with specification of time horizons) – current assessments of employee competencies in comparison with those fixed in item 1 – the assessments of the results of work performed by employees; – plans for hiring, training, firing

[a]Hereinafter, the expression "there is a description" means the presence of a regulated, relevant and substantive specification, fixed in electronic form and used in the course of implementation of activity

5.5   Model of the Process of Executing a Complex Activity

In the previous sections of this chapter, the process models of individual CA elements are presented. The subject of this section is a description of the model of the process of implementing the CA as a system—an integral and interrelated set of its elements.

According to the definition of the system used in the work (see footnote 3), the construction of such a model requires the analysis of three components: the elements of the system, their connections and goal setting.

Analysis of CA elements. A brief review of all the above CA models makes it possible to note that in them:

- the model of the complex activity element is defined (SEA) (Sect. 3.1);
- structural models are considered as a means of integrating typical elements in order to obtain synthetic objects—complex elements of CA (Sects. 3.2 and 3.3);
- fractal and recursive properties of CA models confirm the absence of restrictions on the coverage of simulated phenomena (Chap. 3, Sects. 3.2 and 3.3);
- uncertainty of the behavior of CA elements is a consequence of the occurrence of events external to the elements of uncertainty;
- process models allow describing the existence of CA in time (Sects. 5.1–5.4), formalisms of generating SEAs and creating new technologies provide opportunities for modeling the development of CA as a system (Chap. 4).

As a result, it can be concluded that all SEAs have a single set of system-wide structural and behavioral properties, which are illustrated in Fig. 5.16.

Despite the hierarchical relationships in the logical and cause-effect structures, all SEAs should be considered as **one-type** to within specific features. This means that complex activity is described by a set of the same type of model elements—SEAs. SEAs describe the integration of activity, the formation of complex activity from its elements, that is, in fact, the organizational and management aspects of the CA. Directly the activity resulting in concrete results is modeled by specific elementary operations that are considered as constituent parts of the corresponding SEAs.

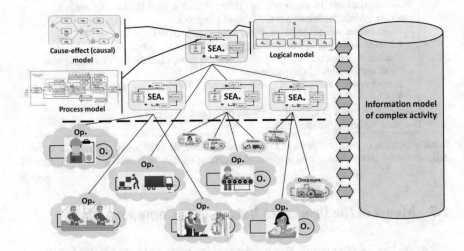

Fig. 5.16 System-wide structural and behavioral properties of SEAs

During the life cycles of SEA, various resources can be used to implement technologies, organize subjects, and form objects of CA. However, an important system-wide resource used by all SEAs without exception is the CA *information model* containing descriptions of SEA technologies, operational information, as well as other information objects specifying CAs (the information model of the CA is dealt with in Sect. 2.6).

Analysis of the links between the elements of the CA (as well as with the external environment) is expedient to perform using the interface widely used in system engineering [62, 136, 137], which is defined in [136] as the *"point of contact"* of interacting elements of the system. In Sect. 3.3, it was determined that the decomposition of the CA goals is the basis for the formation of the CA structure of the SEAs logical structure. This structure allows you to enumerate and pro-analyze all possible options for relationships, links and interactions between elements of the CA. The options are illustrated in Fig. 5.17, where the vertices of the logical structure of the SEA, solid direct arrows—the relationship "goal-sub-goal" and "higher-level SEA—lower SEA", dashed and dotted arrows—the direction of information exchange between SEAs.

Variant 1 of the connection between the elements of the CA, which are in direct relations "downstream-superior", for example, in Fig. 5.17 between *s* and *b*, or *z* and *d*, or *v* and *d*. At different phases of the life cycle of CA, the interaction between such elements occurs in different ways:

- During the design phase, the subject of the higher SEA generates demand, initiates the formation of the subject of the lower SEA and the actual realization of the activity of the lower SEA (the procedures of the PRROR and the PDG in the form of elementary operations 9-a and 10 in Fig. 5.4).

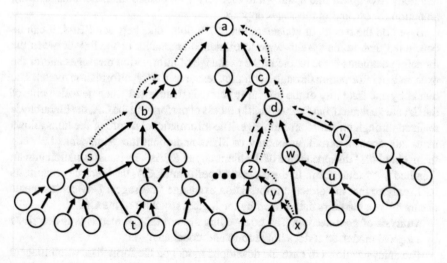

Fig. 5.17 Options for relationships, links and interactions between elements of the CA

- During the technological phase, the subject of the higher SEA initiates the implementation of the actions of the lower SEAs (elementary operations 11-a and 11-b
  in Fig. 5.4).
- During the corresponding phase of reflection, the subject of the superior SEA can
  request and receive information from the lower EDC.
- In parallel, during all phases, the subject of the lower-level SEAs can escalate the
  problems arising in connection with the events of uncertainty to the higher SEAs
  (elementary operation 14 in Fig. 5.4).

The system-wide factor is that in all cases, "direct communication" is realized
in the form of information exchange: either in the form of messaging (bidirectional
dashed arrows in Fig. 5.17), or in the form of reading and writing information into
the CA information model.

Option 2 "indirect" links between elements that are the immediate downstream
elements of one superior SEA, for example, in Fig. 5.17 between the elements z,
w, v and d. Interaction between the elements that are in this connection during the
design and reflection phases is absent. On the technological phase, they indirectly
interact: the subject of the higher SEA synchronizes the execution of such elements
to comply with the cause-effect structure of the superior SEA. In general, such CA
elements access to the same objects of the information model, so we can say that
they exchange information through the information model.

Variant 3 links between elements not in the relations listed above, for example, s
and a or x and d. Elements of CA, which are in "long-distance communication", on
the system-wide level are rather weakly connected. First, their interaction includes
the possible exchange of data through an information model. Secondly, in the case
when one of the elements is superior to another, but through several levels of the
hierarchy (example x and d, shown in Fig. 5.17 with dashed unidirectional arrows),
an option for escalating messages appears.

Based on the results of consideration of possible links between SEAs, it can be
concluded that, at the system-wide level, the commonality of the link between the
modeling elements is the nature of the exchange of information messages and/or the
exchange of information through a common resource, the CA information model. The
hierarchy and fractality of the logical structure is manifested when new elements of
the CA are generated, but further, in the process of performing the CA, the hierarchy is
realized through information exchange. The informational nature of the links allows
us to talk about a certain autonomy of SEAs and about the "soft" links between
them, about the "disappearance of the hierarchy of SEAs", its transformation into an
"implied", "virtual" form. In the process of performing CA, the hierarchy manifests
itself only in that the directed information exchange (messages) between elements
occurs only in the pairs determined by the logical structure of the CA.

Analysis of goal-setting. The goal-setting system for CA elements is described
by a logical model, so it does not require additional analysis.

The analysis allows to base the developed model on the formalism of multiagent
systems [59, 96, 120] and expand it with the properties necessary for a unified
description of the process of realizing any sets of CA elements (Fig. 5.18, on which

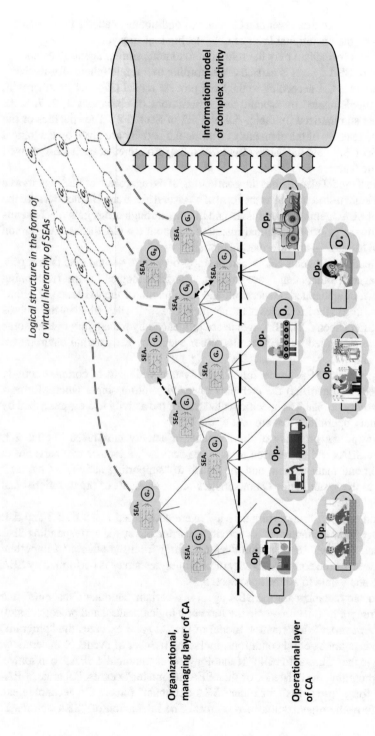

Fig. 5.18 Execution of a set of SEAs

the organizational, control layer of a CA can be conditionally called a layer business processes, and the operational layer of CA—technological layer).

SEAs in this formalism play the role of autonomous similar agents. The change in the status of SEA agents is carried out according to a single scheme for realizing the life cycle of a CA according to the SEA process model (Fig. 5.4 together with Table 5.1, supplemented by specific text descriptions of operations 1, 2, 7, 8, 12 and the rules summarized in Tables 5.2 and 5.3 of Sect. 5.2). The specifics of the agents of the features of the elements of a particular activity are given by the logical structure (Sect. 3.3) and the cause-effect (Sect. 3.4) structures of each SEA, as well as its specific features.

The logical model also defines the goal setting of the aggregate of SEA agents as a whole. The logical structure takes the form of a "virtual hierarchy" in the course of the execution of SEA agents that are linked and interact through messaging mechanisms and normative, a priori, operational information about the CA through a common information resource—the information model.

The collection of agents-SEAs can develop over time—some agents can generate new agents, others can cease to exist, new technologies for the functioning of agents can be generated. Generation and termination of the existence of agents occurs according to the life cycles of the elements of activity, described by the process models of SEA (Sect. 5.2). SEA agents are influenced by uncertainty events, some of which are generated by SEA agents, other events by the external environment (Sects. 2.7 and 4.5).

The aggregate of SEA agents reflects the system-wide part of complex activity that "connects" (organizes) the elements of activity into a single whole, forms a complex activity proper. The specific realization of the activity itself is described by the elementary operations that are part of the SEA.

According to the grouping of the elements of activity introduced in Sect. 2.1, some of the SEAs can be referred to as "core activity", the other two parts are to organization and management, and the fourth to "supporting activity". Chapter 7 is devoted to the system-wide characteristics of the elements of organizational and management activity.

The main types of "supporting activity" were considered in Sects. 5.3 and 5.4: the creation of new technologies for activity is carried out by the corresponding SEA agents (Sect. 5.3), whose execution does not differ from the others; an important practical aspect of the complex activity of providing resources is performed by SEA agents also equivalent to all the rest (Sect. 5.4).

An adequate metaphor for the SEA agent is a certain "automatic machine" that reads the "program" (technology in the form of a logical, causal and process model) from the "*repository*" (information model of the CA) and executes the "program" taking into account external conditions, including number of events of uncertainty. According to the "SEA-automatic" technology, other "automatic SEAs" can spawn, after the "program" the existence of the "SEA-automaton" ceases. Separate "SEA-automata" form "programs" for other "SEA-automata" (create CA technologies), others perform the organization of resources. The interaction of "SEA-automata" occurs through the exchange of messages, as well as the reading/writing of normative, a priori, operational information from/to the "vault".

A dynamically changing set of equally organized, but realizing different "programs" of "SEA-automata" describes complex activity in general, primarily its organizational and managing "layer" of CA. The direct realization of the activity is modeled by specific elementary operations "operating machines", each of which is associated with the "SEA-automaton".

Thus, Fig. 5.18 and the above considerations illustrate the following statement (see also Sect. 7.1).

Multiagent representation of complex activity.

Any complex activity can be represented as an extended multi-agent model.

Summary of This Chapter

Process models of the structural elements of the activity, models for realizing their life cycles are introduced, as well as the models representing the evolution of CA elements, their behavior over time, meeting the requirements (e) … (h) of the MCA, formulated in Sect. 2.10.

Process models together with structural models and generation models form the core of the methodology of complex activity and a typical description of the CA, which is an analog of the architectural template.

Life cycles of different SEAs are identical—they consist of the same phases, stages and stages; and therefore the various elements of activity have the same procedural components. Therefore, the structure of the process model in the BPMN-notation (Fig. 5.4) is the same for different SEAs, it is system-wide. The specificity of SEAs is in their target structure and technology.

Process models of typical elements of activity—goal-setting and creation of new technologies, as well as realization of the life cycle of resources are given.

The model of realization of complex activity as a system is presented, the possibility is shown to base the formalism of multiagent systems on it, expanding it with the properties necessary for a unified description of the process of realizing any sets of CA elements.

Chapter 6
Effectiveness and Efficiency of Complex Activity

This chapter is devoted to the categories of activity evaluation, considered as the process of obtaining the result. The main summary characteristics of such a process are the answers to the questions: what is received, at what time and what are the costs that caused the completion of the process. Formalized, numerical answers to these questions give the values of the corresponding summary characteristics of the CA of its effectiveness and efficiency, which are discussed below.

The activity evaluation is carried out, firstly, in the form of forecasting—a priori, before the activity starts, when creating a CA technology (Sect. 5.3)—to select the best option for CA realization; secondly, a posteriori, upon completion of the actions and evaluation of the result (Sect. 5.2). Any prediction is hampered by the uncertainty of the future behavior of the object in question. Therefore, the task of predictive evaluation should be structured:

(a) it is necessary to identify the factors of complex activity affecting the characteristics of its effectiveness and efficiency;
(b) predict the change in these factors;
(c) based on this, build a forecast of effectiveness and efficiency.

The forecasting procedures are specific and therefore are beyond the scope of this work. However, the factors themselves and the characteristics of efficiency and effectiveness differ in systemic generality and are therefore considered in this chapter.

In the first of the three sections of this chapter, system-wide properties are analyzed and identified, and a formalism is proposed for describing the result of CA. In the second section, a brief review of known studies devoted to the efficiency of complex systems and related concepts is made. Based on it, definitions of performance characteristics and efficiency of complex activity are formulated and corresponding general quantitative indicators are proposed. The third section is devoted to identifying and discussing the factors of complex activity affecting its effectiveness and efficiency.

M. V. Belov and D. A. Novikov, *Methodology of Complex Activity*, Studies in Systems,
Decision and Control 300, https://doi.org/10.1007/978-3-030-48610-5_6

6.1 Model of the Results of Complex Activity

Despite the complexity of the goals, the subject and result of the CA, the process of obtaining the result in the course of complex activity and correlating it with the goal has system-wide properties characteristic of any CA, regardless of its industry specificity. Let us list these properties, analyzing the evolution of the object in the course of the CA realization.

The *subject of the activity* is described by a set of characteristics that take on certain values at each moment (Fig. 6.1). From this point of view, activity is a process aimed at changing the characteristics of the subject of activity, translating the values of these characteristics from the initial values to the target ones.

The *result* of the activity is the terminal state of the object in the process of its change in the course of the realization of the activity, the state of the object at the time of the completion of the activity. Specification of the *goal* determines the desired result desired, anticipated or objective state of the object, that is, the desired or target values of the characteristics of the object at the time of completion of the activity. Obviously, to achieve complex goals, it is necessary to organize and implement complex (adequately complex) activity.

It is important to note that the achievement of the goal (bringing the characteristics of the subject to the target values) should be evaluated from the point of view of meeting the *demand* external to the activity and the *need* corresponding to it. In the

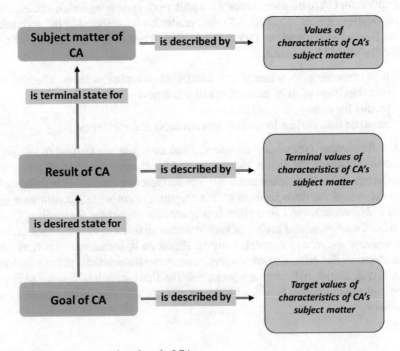

Fig. 6.1 Subject matter, result and goal of CA

case of rational activity, the achievement of a goal is necessary for the subject of activity only insofar as it corresponds to some external demand for him, which he "accepted", and this became his internal need.

The satisfaction of the need is traditionally characterized by *utility*, so we use this category, and we will talk about the utility of the result in terms of external demand.

Classify the elements of the CA, depending on the result and its usefulness for two reasons.

First, the SEA result can be characterized by one of two alternative methods:

a. The fact of obtaining the result of the CA, "partially" the corresponding goal, is indistinguishable from the fact of the lack of activity, and the various outcome options that meet the objectives are equivalent to each other. In this case, the set of values of the utility function consists of two elements, which can be called "goal achieved" and "goal not achieved." We call this result (and subject) conditionally *"binary"*.

b. Different values of the characteristics of the object differ from each other the results described by them, then the set of values of the utility function includes more than two elements. This result (and subject) of the CA will be conditionally called *"continuous"*.

Secondly, the CA can end with one of the alternative outcomes:

i. When the continuation of the CA of a superior SEAs is made without the repetition of this CA, given by the SEAs, or the result of this SEAs does not require the repetition of activity in terms of an external source of demand.

ii. When from the point of view of the activity of a superior SEA or an external source of demand, it is necessary to repeat the CA of this SEA or a significant change in the technology of this and/or superior SEA, or to refuse to perform activity due to the realization that it is impossible to achieve the goal.

In the case of a binary SEA result, the set of possible values of the characteristics of an object is naturally divided into two disjoint regions, "target", when the goal is achieved, and all the rest, when the goal is not achieved.

In the case of "b", outcomes i and ii divide the entire set of possible values of the characteristics of the subject into the "target area" (outcome i), if the characteristics of the result fall into which one can say that the result corresponds to the demand/demand, and the rest, when the result does not correspond to the target parameters.

Summarizing these cases, we will define the areas of possible values of the characteristics of the SEA results as the "goal is achieved" area (the target area) and the "goal is not achieved" area. In the "target not achieved" area, two corresponding subregions can be identified (Fig. 6.2).

Elements of the "target area" can be indistinguishable from the point of view of utility for the case of a binary result, or have different utility in the case of a "continuous" result.

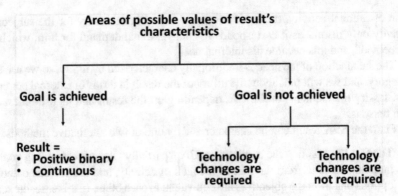

Fig. 6.2 Domains of feasible values of CA result characteristics

The binary result is typical for the overwhelming majority of CA elements of design types, "design SEA", but can also occur for process elements of CA; "continuous" result is possible for both project and "process SEA".

For example, the execution of such elements of activity as "processing documents and issuing a loan to a retail bank customer" or "conducting routine maintenance work on certain equipment of a power unit" can be completed only with a binary result, while "generation of electricity during the day" has "continuous" result.

The introduction of two alternative areas on the set of values of the characteristics of the result and the object of the CA makes it possible to determine the following enlarged states of the SEA:

A. "Implementation of activity": the goal has not been achieved at the time specified by technology and the resource consumption did not exceed the level specified by the technology (activity is not started, as a special case). This means that the CA element described by the SEAs is not implemented, and the CA can be continued as an execution of these SEAs.
B. "The goal is achieved" for no more than the time specified by the technology and the resource consumption did not exceed the level specified by the technology. This means that the CA element described by the SEA is successfully implemented, the CA can be continued.
C. "The goal is not achieved" for more than the time specified by the technology or the actual resource consumption exceeded the planned one. This means that the CA technology must be changed. That is, the initial CA is terminated and a new CA is started (a special case is the refusal of activity).

A generalized diagram of states and transitions between them can be represented in the form Fig. 6.3.

The enlarged state of the SEA is used to determine and analyze performance indicators and the effectiveness of complex activity.

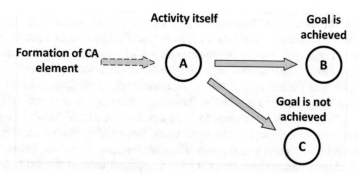

Fig. 6.3 The enlarged state of the SEA and transitions between them

6.2 Definitions and Criteria of the Effectiveness and Efficiency of Complex Activity

Define the indicators and criteria associated with the *effectiveness* of complex activity. Efficiency in one way or another includes such related concepts as effect, value, result, effectiveness. All of them are intuitively understandable and widely used in both practical and scientific spheres—economics, management, systems theory, system engineering, engineering disciplines and other fields. Despite their wide prevalence, starting, for example, with classical works on the theory of systems (see [14], etc.), there are no generally accepted definitions of these concepts and their interrelations. Therefore, we will briefly analyze current sources and, on this basis, formulate appropriate definitions.

Among the works devoted to the effectiveness of complex systems, it is necessary to highlight a fundamental collection of articles [5], whose goal is to consider the concept and methods of evaluating the effectiveness of complex systems by "non-monetary indicators". The collection includes articles devoted to the history of philosophical concepts of "cause" and "effect", system-engineering views on the problem, examples from the socio-cultural sphere, ecology, education and others. Of greatest interest is the paper [93], which examines efficiency from the point of view of the structure of the system, the measure of efficiency, the resources expended, and the *value* created, taking into account both technical and socio-economic aspects. Value is understood in a broad sense, including both the measure of productivity, and the measure of utility, and broader philosophical interpretations.

You can also consider vector effect indicators: the cumulative effect should include its components: scientific and technical, economic and social effects, ecology. In general, the effectiveness of activity or a project, including innovative ones, is a category that expresses the correspondence of the results and costs to the goals and interests of their participants.

A popular method for studying the effectiveness of complex systems is the *Data Envelopment Analysis* (DEA). This method was first proposed in [26] on the basis of ideas [43]. Currently, the number of international publications on this topic is several thousand. The method of DEA arose as a generalization of simple indicators to the case when the behavior of a complex object is described by a set of input parameters and a set of output parameters. For the correctness and meaningfulness of such a formulation, as well as for eliminating subjective factors in models, a number of similar complex objects are considered within the framework of the DEA. The method of DEA has a deep connection with the theoretical economy, system analysis, multi-criteria optimization. It allows you to build a multidimensional "economic space", find the optimal trajectories, calculate the quantitative and qualitative characteristics of the behavior of objects and simulate different situations. However, despite practical applicability, this method does not have a sufficient level of generality from the theoretical point of view.

The following definitions are possible:

- *effectiveness*—a relative indicator characterizing the degree of achievement by an economic entity of the set goal in one and/or several spheres and all its economic activity at a certain point in time;
- *efficiency*—a relative indicator characterizing the positive dynamics of the development of a subject of the economy at a certain point in time and equal to the ratio of the result (effect) to the costs that caused its receipt.

This pair of indicators assesses, first, what was obtained during the realization of the CA, and secondly, what are the costs at the same time. On their basis, we define the corresponding categories for the CA and, supplementing them with a time factor, we introduce the numerical values of the CA.

The *effectiveness* of the execution of the element of complex activity, SEA—the degree of correspondence of the result to the goals of the element of CA, SEA.

The *efficiency* of the execution of the element of complex activity, SEA—the degree of correlation between the result obtained and the characteristics of resources used in the course of CA realization.

Using the adopted management definitions (see footnote on p. 14), we formulate similar categories for management and organization:

The *effectiveness of the SEA/a management/organization* is by definition equal to the effectiveness of this CA element (since the control objective is the corresponding change in the states of the controlled CA element).

Efficiency of management/organization SEA/a—the degree of correlation between the result of this element of CA and the characteristics of resources used in the management/organization of its activity.

In the most general case, these definitions set the same general quantitative indicators:

- *effectiveness* as a function (for example, utility function), defined on the set of possible values of the characteristics of the result (depending, in general, on the time) and the time of obtaining the result,
- the *efficiency* of a function defined on the set of possible values of the characteristics of the result, the time it was obtained, and the characteristics of the resources, which also generally depend on time.

The introduced indicators allow you to build on their basis the criteria by which you can compare and choose (optimize) the various options for implementing the elements of activity, options for organization and management. To do this, one of the indicators is constrained, and the other is declared to be the target, and its value compares the options for implementing the elements of the CA.

Performance and efficiency functions are usually defined in fairly simple forms that allow constructing transparent criteria, and, at the same time, reflecting the meaningfulness of comparing and optimizing the execution of CA elements.

In this section, it is appropriate to mention the *problem of aggregating* character-istics and performance indicators. Indeed, even the characteristics of a single SEA can represent a vector, not to mention the characteristics of a multi-level hierarchy of SEAs. The problem of aggregation is how, knowing the characteristics of the activity of elements of some level of its logical structure, find the values (preferably scalar) of the indicators and/or the characteristics of the elements of the next higher level of the hierarchy [21]. A universal answer to this question does not exist, his search for many years involved in the *theory of measurements* [107], *multi-criteria optimization* and many other sections of modern science. In each specific case, the solution of the aggregation problem is specific, the general recommendations are the competent use of *measurement scales* and the corresponding transformations of mea-surement results, as well as monotonicity monitoring—for example, improving the performance of downstream SEAs should not lead to a decrease in the performance of higher-level ones, etc.

6.3   Factors Affecting the Effectiveness and Efficiency of Complex Activity

Having determined the notions of effectiveness and efficiency of an element of com-plex activity, it is expedient to analyze and reveal the factors determining their val-ues. We use for this purpose the SEA model (Sect. 3.1, in general, and Fig. 3.2, in particular), having considered all its component parts.

The target aggregate ("need-goal-task") directly sets the desired shape of the result—the target area of its parameters, and thus affects the performance of the SEA. Since goals and tasks do not determine the way to achieve the goal, including resources, therefore their impact on efficiency is indirect.

Technology (forms, means and methods of activity) determines, first, the possi-bility of achieving the result, that is, the satisfaction of the condition of belonging to the parameters of the target area, and secondly, the area of planned values of the characteristics of resources. Technology generally specifies the dependence (not necessarily functional) of the required characteristics of resources on the character-istics of the result and the time of its achievement, this dependence can in general be described by a set of the above models (Chaps. 3–5), symbolically reflect this dependence by a "technological predicate".

The subject of the activity performs the action, realizing the technology, that is, the "technological predicate". From the point of view of the external observer, first of all, the subject of the higher SEA, the independent influence of the subject (the SEA under consideration) on the result, including the active choice, and therefore on the effectiveness and efficiency, can be described through the manifestation of the subject's uncertainty (Sect. 2.7).

The subject of activity, being a passive component of the SEA, affects the effectiveness and efficiency also through the manifestation of the uncertainty of the subject (Sect. 2.7).

In the process of performing the action, the technology is realized and the result is directly formed, at the same time all types of uncertainty are realized (Sect. 2.7)— subject, technology and external environment.

From the above considerations follows the statement:

Factors determining effectiveness and efficiency of complex activity.

Effectiveness and efficiency are completely determined, first, by the content of the "technological predicate", and secondly, by the values of the characteristics of uncertainty.

The result of the activity depends on the management; therefore, knowing this dependence, and also defining the criteria for the effectiveness and efficiency of the activity, it is possible to calculate the *management effectiveness* as the efficiency of the activity realized in the given management [99]. Having the criterion of management effectiveness, it is possible to set and solve the problems of *management optimization*—the search for an admissible control having maximum efficiency [101].

The assignment of the functions of effectiveness, efficiency and the "technological predicate" allows us to set and constructively solve the formal problems of a priori comparison and optimization of the execution options for CA elements (the efficiency criterion is "functionally constructed" from the performance vector and the performance and effectiveness indicators), but this requires mechanism for eliminating uncertainty in terms of factors that can not be controlled or influenced. To analyze the system-wide properties of such a mechanism, consider the events of uncertainty and their impact on the effectiveness and efficiency of CA.

Uncertainty events cause a reaction that is described by the "chain of reaction to the uncertainty" of the SEA process model (Sect. 5.2), that is, certain fragments of the "technological predicate".

The "chain of uncertainty response" of the regular (Sect. 4.2) and replicative (Sect. 4.3) activity only include elements of escalating problems to the subject of the superior SEA and initiating a priori the described elements of activity. In these cases, the "technological predicate" is completely defined, therefore the characteristics of uncertainty affecting effectiveness and efficiency are only the intensity or the possibility of occurrence of uncertainty events of one type or another, both measurable and true (Sect. 2.7). Other characteristics of uncertainty events are already taken into account in the content of the "technological predicate"—in the rules of the event specification (Sect. 5.2).

The "chain of response to the uncertainty" of creative activity (Sect. 4.4), in addition to the above-mentioned options for escalating and initiating known elements, also includes elements for the creation of new CA technologies. In this case, the "technological predicate" contains in the general case a priori completely unknown fragments: SEAs, which can arise as a result of the synthesis of new technologies. These fragments are generated by events of true uncertainty, with respect to these events, neither the possibility of their occurrence nor other parameters of events is known.

Thus, events of uncertainty can be classified on three grounds:

(a)   the uncertainty is measurable or true;
(b)   the possibility of an event can be described by a probabilistic model, yes/no;
(c)   is the model of other characteristics of events known that is identical to the division into reproductive and productive activity (see Fig. 4.2), and also the completeness of the description of the "technological predicate", yes/no.

The results of classification are presented in Table 6.1.

Measurable uncertainty (case I, first line) is eliminated by applying traditional probabilistic methods—in particular, Bayesian or maximum likelihood (or their variations, for example, sequential ones).

The true uncertainty (Case II, second line) with a known model of event parameters and a completely defined "technological predicate" (the "chain of reaction to uncertainty") can be eliminated by applying expert (or any other) assessments of the possibility of occurrence of events and reduced to case I. Full knowledge of the "technological predicate" means that, despite the lack of knowledge about the possibility of an event, its characteristics are known to such an extent that it is possible to estimate (possibly in the form of a probability Second model) the consequences of its occurrence and, consequently, to determine the response to it. Examples of such events are emergencies arising on aircraft for which the actions of the crew have been

Table 6.1 Classification of events of uncertainty and possible elimination methods

Cases		Classification bases for uncertainty (see above)			Methods to eliminate uncertainty
		a	b	c	
Cases	I	Measurable	Probabilistic	Known	Probabilistic-statistical and analogous models (PSM)
	II	True	Unknown	Known	PSM + expert appraisals of events occurrence
	III	True	Unknown	Unknown	PSM + scenario approaches, which actually replace the study of CA and environment

a priori determined, or project accidents at nuclear power plants and other high-risk facilities.

True uncertainty with completely unknown a priori characteristics of events (case III, third row of Table 6.1) is fundamentally different from the previous case. Such uncertainty means that all or some of the components of the activity (subject, technology or object) and/or the external environment are so poorly understood that the performance of the activity is accompanied (or even actually!) by the acquisition of new knowledge, which leads to a change in technology. In this case, the elimination of uncertainty can be performed only in ways that replace the study of the components of the CA and/or the external environment, for example, the development of expert scenarios and the modeling of CA based on them. The scenario approach is well developed (for example, [34]) and is widely used for forecasting in areas where the true uncertainty is most significant—in the social, political sphere, in the economy. The use of scenario and similar approaches is in fact subjective, not justified, not a scientific way of forming knowledge with all its inherent flaws. However, in conditions when a full-fledged scientific research is impossible for one reason or another, in practice this approach is applied.

It makes sense to note that the aforementioned methods for eliminating uncertainty eliminate both measurable and true uncertainty from the model under consideration, while the influence of uncertainty on the CA remains. Therefore, such techniques are artificial, and as applied to true uncertainty, they are also subjective. These features of eliminating uncertainty must be borne in mind in practical use, applying them with caution and certain conservatism.

Summary of This Chapter
The general system properties of the CA result are analyzed and isolated, and a general formalism of its description is proposed.

Definitions of categories of effectiveness and efficiency of complex activity are formulated and corresponding quantitative indicators are proposed.

Factors influencing the effectiveness and efficiency of complex activity are revealed.

Chapter 7
Organization and Management of Complex Activity

Organization and *management*, on the one hand, are special cases of complex activity, on the other hand they have other elements of complex activity. In the introduction and Sect. 1.1, the concepts of organization and management were introduced and their system-wide aspects were preliminarily examined. In Sect. 1.1, it was noted that any complex activity includes elements related to both organization and management.

After the CA models are introduced in Chaps. 3–5, the concepts of organization and management can be formally analyzed in terms of the proposed models using the "language" of SEA, logical, cause-effect and process models of CA.

The structure of the presentation of the material in this chapter includes five sections. In the first section, the place and role of organization and management in the system of CA elements are considered, and the links between managing and organizational SEAs with other elements of the CA are defined. The second section is devoted to the structure of organization and management. The third section discusses the problem of optimization of complex activity. The fourth section describes the tasks of optimizing the performance of complex activity with known technology. In the fifth—the problem of synthesis of optimal technology; the results of the section directly respond to the requirement i to the methodology of complex activity and support the remaining requirements for it, formulated in Sect. 2.10.

7.1 Place and Role of Organization and Management in a System of Components of Complex Activity

A system-wide analysis of organization and management should begin with the confirmation of the thesis that "any comprehensive activity includes elements related to both organization and management". Above were introduced the definitions of organization as an activity aimed at creating internal orderliness, coherence of interaction of more or less differentiated and autonomous elements; and management as

M. V. Belov and D. A. Novikov, *Methodology of Complex Activity*, Studies in Systems, Decision and Control 300, https://doi.org/10.1007/978-3-030-48610-5_7

an impact on a managed system (management object), designed to ensure its (its) behavior, leading to the achievement of the objectives of the subject.

The consideration of the SEA formalism (Sect. 3.1), the logical model (Sect. 3.3), the causal model (Sect. 3.4), the models of generation and uncertainty of the CA (Chapter 4) does not allow us to uniquely identify elements that are necessarily related to organization and management: any of the elements listed models can both belong to the sets of "elements of organizational activity" and "elements of management activity", and do not belong to them.

The SEA process model (Sect. 5.2) in its most general form describes the fulfillment of the life cycle of a CA element, beginning with the fixation of demand and ending with a reflection, that is, the realization of the entire life cycle of a CA; therefore, it is expedient to consider the process model in more detail.

The first six stages of the LC (Fig. 5.4) are not directly related to the organization of the downstream elements, nor to the influence on their behavior.

Consider the steps 7–9, in Fig. 7.1, a fragment of the SEA process model is shown (Fig. 5.4), including the system-wide elements of the design phase (the stages of the assignment of subjects and resources) of the life cycle of the SEA.

These elements, together with the links in a large format, are shown in the main part of Fig. 7.1, and in the right upper corner of this figure a reduced sketch of the complete cause-and-effect structure of the process model is shown (Fig. 5.4), and on it the marked elements are marked. This graphic format is chosen to focus on the elements of interest on the one hand, and on the other—to see the place and role of these elements in the structure of the activity as a whole.

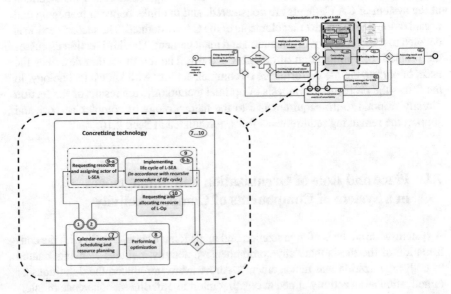

Fig. 7.1 The system-wide elements of the design phase of A-SEA LC

The main element in the design phase is the pair of elementary operations 7 and 8 "Perform calendar and resource planning" and "Perform optimization". The specific form of operations is specific, but as a result, they initiate the following elements of activity:

- formation of subjects of lower-level SEA (9-a);
- recurrent realization of the life cycles of each of the lower-standing SEA (9-b);
- formation of resources for the technology of each of the subordinate operations (10).

Initiation of elements of activity provides the required sequence of their execution, that is, their behavior, therefore, is management.

The formation of subjects and resources is carried out by means of an element of activity on request, obtaining and organizing resources (see Sect. 5.1), within which links are established between the subject of the A-SEA and the subjects of the L-SEA and between the subject A-SEA and the resources of L-Ops. Therefore, the fragment in question is an element of organizational activity, which is carried out in the realization of any complex activity.

Similarly, the corresponding fragment of the execution phase of the SEA life cycle (Fig. 7.2) and the chain of uncertainty response can be considered (Fig. 7.3).

In Fig. 7.2, operation 11-c "Controlling the fulfillment of conditions" implements the control activity: it checks the preconditions-vertices of the cause-effect structure

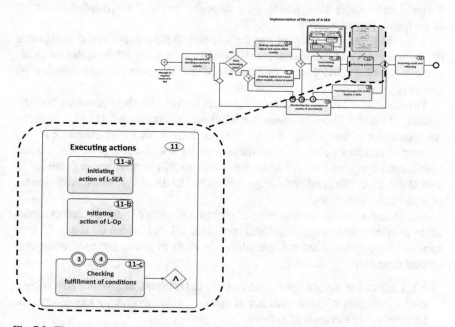

Fig. 7.2 The system-wide elements of the implementation phase of A-SEA LC

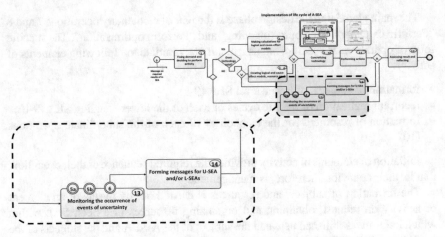

Fig. 7.3 The system-wide elements of the response to uncertainty in A-SEA LC

and, through operations 11-a and 11-b, directly initiates the execution of SEAs and operations.

The fragment of the chain of response to uncertainty (Fig. 7.3) also consists of elements of management and organizational activity: during the operation "Monitor the occurrence of uncertainty events", the occurrence of uncertainty events is verified, the problem escalates through event 14 or through event 5—a return to the re-creation of the technology of activity.

Thus, the above analysis confirms the presence in the process model of elements of organizational and managerial activity and, thus, formally substantiates the thesis that any complex activity includes elements of the CA, related to both organization and management.

The analysis of the process model allows us to conclude that operations "Initiate actions" (11-a and 11-b), "Monitor the fulfillment of conditions" (11-in), "Monitor the occurrence of uncertainty events" (13) are carried out in the execution of any element of complex activity, "Implement calendar and resource planning" (7) and "Implement optimization" (8) which are elements of the management activity, as well as the SEAs "Request and Assign Resources" (9-a and 10), which is an element of organizational activity.

In conclusion of this section, we will outline a number of general characteristics of the management and organizational operations X1–X3, which we use in the next section. Comparison of control operations allows us to generalize and isolate their general characteristics X1–X2:

X1. Each of the control operations consists in (constant, or repeated, or continuous) verification of the occurrence of certain conditions and the initiation of the corresponding elements of activity.

X2. Both the conditions and the triggered elements of activity are specified by logical, cause-effect and process-based models of CA, that is, actually system-wide CA technology.

At the same time, the content of the organizational element "Request and Assign Resources" (9-a and 10) is to obtain resources and establish links with them. This element is performed in two cases: (1) for the purpose of providing resources to the technology of the downstream operation (10) or (2) to determine the subject of the lower-level SEA (9-a). In the first case, a connection is established between the subject of the A-SEA and the resources of the subordinate operations, in the second case, between the subject of the A-SEA and the subject of the lower SEA. This allows us to formulate the characteristics of the element of organizational activity X3:

X3. Each element of organizational activity ensures the request and receipt of resources, and the establishment of links between the higher-level entity and them.

7.2 Management and Organization Structure

Consider the activity of the subject in the course of implementing the life cycle of the A-SEA (see also Table 7.1 and Fig. 7.4):

- At the first stage of the design phase, during the fixation of demand and awareness of the need, the potential subject of the A-SEA *analyzes* the demand and the situation as a whole, the experience of its previous activity and its general opportunities to meet demand; in the end, he decides to carry out activity. The activity of the subject as a whole is of the nature of analysis (decision-making on the basis of analysis is a one-step act), there are no downstream elements of activity at this stage.
- In the stages from the second to the sixth design phase, the A-SEA subject creates a logical and cause-effect model, forms resources, that is, *synthesizes* the future activity (its and L-SEAs), its elements, the relationships between them, etc. in the form of models. Subordinate elements of activity are created in the form of information models, neither the subjects, nor the resources of these elements have yet been created.
- In the stages from the seventh to the tenth phase of the design, the causal model is first specified in the form of calendar-network plans/schedules, and then the request and receipt of resources for the designation of subjects of lower-level SEAs and the provision of technologies for subordinate operations-filling the roles of subjects and resources with specific instances. Thus, the establishment of specific links between the subjects of higher and lower SEAs with each other and with resources, i.e. is the *concretization* of activity. At these stages, in addition to the subject A-SEA, subjects of L-SEAs are created and L-Ops resources are formed.

Table 7.1 Phases, stages and steps of life cycle and activity of A-SEA

Phase	Stage	Step	Activity of subject
Design	I. Fixing demand and understanding needs	1. Fixing demand and understanding needs	Analysis of: demand, opportunities, external conditions and previous activity
	II. Setting goals, structuring goals and tasks	2. Creating logical model	Synthesis of activity
	III. Selecting and developing technology	3. Checking the readiness of technology and the sufficiency of resources	
		4. Creating cause-effect model	
		5. Creating technology of lower-level elements	
		6. Forming/modernizing resources	
		7. Calendar-network scheduling and resource planning	Concretization of activity
		8. Performing optimization	
		9. Assigning actors of L-SEAs and defining responsibilities	
		10. Allocating resources for L-Op	
Implementation	IV. Performing actions and obtaining results	11. Performing actions and obtaining results	1. Execution of actions 2. Regulation of activity
Reflection	V. Assessing results and reflecting	12. Assessing results and reflecting	Assessment of activity

- At the eleventh stage of the execution phase, the subject of the A-SEA, firstly, being the subject of the elementary operations of L-Ops that are members of the A-SEA, directly *performs/executes* the operations (the right branch in Fig. 7.4). Second, also at the eleventh stage it controls the onset of conditions according to the technology of activity and initiates the actions of the children of the CA (and L-SEAs and L-Ops), that is, *regulates* (the left branch in Fig. 7.4) the activity of the subordinate elements of the CA and the elementary operations for which it himself acts as a subject (then it is a *self-regulation*). Regulation consists in making decisions within the framework of a given technology of activity, including in responding to uncertainty events. At this stage, naturally, in addition to the A-SEA, all the downstream elements of the L-SEA and L-Op are involved.

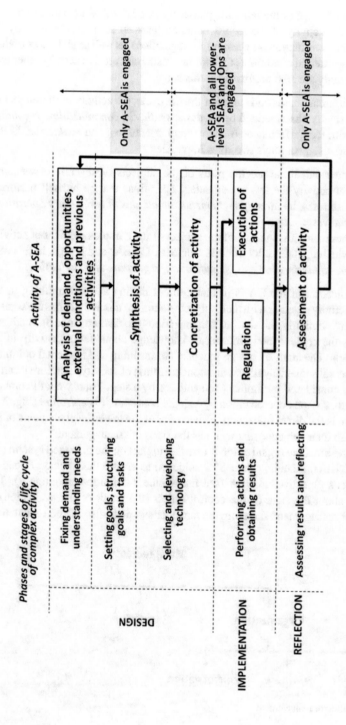

Fig. 7.4 The activity of the subject in the course of implementing the life cycle of CA

- At the twelfth stage of the reflexion phase, the A-SEA subject performs an *evaluation* of the activity of the results obtained (as well as those obtained by the L-SEAs, this information is necessary for regulation), as well as all factors of their obtaining—external conditions, technology, resources, etc. No activity other than A-SEA activity are also performed at this stage.

Thus, at different phases and stages of the life cycle, the activity of the subject A-SEA is consistently characterized by *analysis, synthesis, concretization, regulation* (in parallel with the performance of elementary operations) and *evaluation*. In this case, the A-SEA subject plays two roles in parallel:

- On the one hand, these are the roles of the subjects of several "elementary" elements of activity-the lower elementary operations that he himself performs. Within these roles, he *directly implements the actions of elementary operations, directly receives results*.
- On the other hand, it is the role of the subject of the complex element of activity, A-SEA, which includes a lot of child elements, L-SEAs and L-Ops. In this role, *he performs analysis, synthesis, specification, regulation and evaluation*.

The specification of the CA is in fact a direct linking and streamlining of the elements of activity together with their subjects, therefore this activity has the signs of "organization". Similarly, the signs of "organization" are distinguished by synthesis: its content is the creation and ordering of knowledge about future activity in the form of models. The analysis has no independent meaning and is carried out in the interests of the subsequent synthesis and concretization of activity, therefore it can be conditionally considered preliminary to synthesis by a step. Thus, we will combine analysis, synthesis and concretization by a single concept of *organization* (Fig. 7.5).

Regulation is the direct impact on the behavior of the lower-level elements of the CA along with their subjects, that is, it has the signs of "management."

At the same time, the organization is also "an impact on the managed system (the entire set of SEAs, including their STS subjects) to ensure their behavior leading to the subject's objective" A-SEA (see management definition, footnote 7). The information about the state of the control object (the result of the evaluation) is necessary for management, therefore the management must also include evaluation.

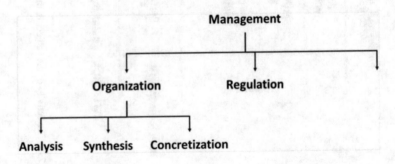

Fig. 7.5 Structure of management

Therefore, it makes sense to say that *management* includes *organization*, *regulation* and *evaluation*. Management in a narrow sense can only be considered as regulation. The object of management (and organization) is the A-SEA and all the downstream elements of the CA, including their subjects—STS.

Thus, we formulate the assertion:

Components of organization and management

Management components are organization, regulation and evaluation. The components of the organization are analysis, synthesis and specification.

The arrows in Fig. 7.4 reflect the logical sequence of component execution. The reverse arrow from assessment to analysis reflects possible cyclical repetitions of activity with "*self-management/self-regulation*" by rethinking needs, setting goals and changing technology of activity, etc. (see also Fig. 1.2).

Management components are always implemented consistently, forming, on the one hand, the life cycle of an element of complex activity, and on the other hand, a *management cycle* that generalizes the Fayol and Deming cycles and the like.

It should be noted that the model of the life cycle of an element of complex activity and the activity of its subject (Fig. 7.4) is applicable to any control systems (see Table 2.5): and to (a) *active system* or (b) man-machine, *ergatic systems*, when the subject (in the particular case of an individual), and to (c) *automatic control systems* (ACS), when direct control (regulation) is carried out by an automatic device that does not contain people—the "control automaton". All the previous presentation of this paper is devoted to the consideration of elements of CA, in which the role of the subject is played by STS; hence, it covers cases (a) and (b). Applicability of the model Fig. 7.4 to case (c)—ACS—follows from the obvious fact that any control automaton is (for the time being, while creative activity remains the prerogative of a person) the subject and result of some complex activity. Also for ACS, the life cycle can be considered as an SEAs for which there exists a subject in the form of STS. In the course of execution of the LC of the control automaton, this entity directly realizes:

- analysis (requirements for ACS);
- synthesis (ACS, including control algorithm, design, device manufacturing, testing, etc.);
- specification (integration of ACS into a managed system);
- evaluation (the results of the ACS operation).

The application of the control automaton, when it implements automatic control, can be considered as an indirect fulfillment of the activity of regulation by the subject.

The components of activity that are variable and unchanged (as a result of the corresponding activity of the subject) are given in Table 7.2.

Table 7.2 Variable and invariable components of activity

Are changed (for A-SEA and lower SEAs)	Are NOT changed (for A-SEA and lower SEAs)
In the course and result of synthesis by A-SEA	
Needs of A-SEA; Goals/tasks of A-SEA; Technologies of A-SEA	Actor of A-SEA; Actions of A-SEA; Subject matter of A-SEA L-SEAs, L-Op
In the course and result of activity concretization by A-SEA	
Subjects; Needs; Goals/tasks; {the sets of L-SEAs}; {the sets of L-Ops}; technologies	Actions; Subject matters
In the course and result of regulation by A-SEA	
Actions of L-SEA; Subject matters of A-SEA; {the sets of L-SEAs}; {the sets of L-Ops}; Actions of L-SEAs; Subject matters of L-SEAs	Actor of A-SEA; Goals of A-SEA; Technologies of A-SEA

Thus, **all the complex activity, except the execution of elementary operations, is management**, and in the latter "creative" (non-routine) is the activity of analysis and synthesis of activity. The most essential elements of the CA, related to analysis or synthesis, are creative (see Sect. 4.4), and they are, to the greatest extent, characterized by true uncertainty. Therefore, on the one hand, they are the most "complex", generate the main share of problems and require the greatest expenditure of resources. But on the other hand, these elements of the CA are the basis for the development of activity and civilization in general. Projecting activity, especially synthesis, generates new creative SEAs in the general case. They create, including technologies and resource pools, their life cycles also include their phases of design, implementation/execution and reflection: a general description of the structure of the LC is fractal (see Fig. 7.7).

At the same time, the concretization is performed according to the CA technology, i.e. links correspond to: goals ("subject-goal" communication), technologies ("subject-technology" communication), hierarchy (subject-subject communication), sub-goals/tasks ("goal-sub-goal"), resources (communication "subject"—"resources"), results ("subject-result" communication), etc. Relationships are relationships between elements (elements of activity, their components, including components of different elements of activity); communication can be different—cause and effect, information; in the general case, the relationship of the relationship of responsibility, subordination, and so on.

The implementation phase of the regulation of activity includes the administrative impacts of higher SEAs, the choice of actions by subordinates, etc., is in many respects "mechanistic", it is regulated by technology, as well as specification.

Figuratively speaking, analysis and synthesis is the development of a "program" (technology), the concretization is the binding of the program to calendar time and allocation of resources for it, and regulation is "the interpretation of the program (technology)." Then we can conclude that the very realization of the activity (as well as the specification) is "simple and automatic". All problems and achievements lie in the creation of technology ("writing programs"). The effectiveness of obtaining the result of the program is determined by the program itself (technology) and malfunctions (uncertainty), as well as the ability to react effectively to unknown a priori failures (true uncertainty) as a result of reflexive control—to generate a new program (CA technology).

In practice, managers perform all types of activity: analysis, synthesis, specification, regulation and evaluation, probably not dividing them and not thinking about it. This creates a kind of "mystic governance," the thesis of the impossibility of formalizing it, etc. And in theory, the whole modern science of management "kneads" the management of the creation of both technology and resource pools. In fact, even the management and organization of the process of creating a new technology is also routine. The activity of the one who creates the technology is creative and complex. It is in this activity that the true uncertainty manifests itself, to which it is necessary to react constructively.

All *management mechanisms* studied within the framework of the management theory of organizational systems [21, 86, 101], are "organizational", since prescribe which management decisions in the implementation phase should be taken by the SEAs in a given situation (depending on the state of the environment, the behavior of the lower subordinated SEA, etc.).

Based on the statement about the components of the CA, the statements and the arguments obtained in the previous sections, we formulate a number of important statements.

Section 5.5 shows that CA as a whole is realized as a set of SEAs and specific elementary operations; let's call this statement the thesis **T1**.

In Sect. 5.2 it is noted that the performance of any element of the CA, and therefore of any system of CA elements, that is, a CA as a whole, is realized by a set of elementary operations and SEAs, presented in Fig. 5.4, which includes (let's call it the **T2** thesis):

- specific operations (1, 2-a, 7, 8, 12),
- the above system-wide elementary operations 11-a, 11-b, 11-in, 13, and SEAs 10,
- also the set of these operations and SEA itself (through a recurrent reference to SEA 9b).

The consolidation of theses T1 and T2, the characteristics of X1, X2 and X3 (see the conclusion of Sect. 7.1) and the statement "On the Components of Organization and Management" allows us to formulate the following statement about complex activity as a complex system and system consisting of systems.

Composition of complex activity

Any complex activity (as well as any of its elements that are complex activity) can be represented in the form of an aggregate of elements of the following types organized and united by a common goal-setting:

- specific elementary operations (representing elementary activity);
- control elements of activity (SEAs or elementary operations) that implement analysis, synthesis, specification, regulation and evaluation of CA.

The links between the elements (their organization) are the nature of the exchange of information messages and/or the exchange of information through a common resource—the CA information model.

The most "complex" components of management are analysis and synthesis. While the specification (consisting in establishing and maintaining links between the subjects of CA elements, resources and subjects of the subordinate elements) and regulation (constant or multiple, or continuous verification of the occurrence of certain conditions and the initiation of relevant elements of activity) are "simple" and routine.

The statement about the composition of the CA justifies the correctness of the statement about the *multiagent representation of the CA* proposed in Sect. 5.5 and illustrated in Fig. 5.18.

From the statement about the composition of the CA it follows that the SEA process model (supplemented by the goal-setting and technology creation model and the model of the resource pool) fully describes the generalized management of complex activity—some "universal control algorithm for CA":

Universal algorithm for a control of complex activity

The management of any element of complex activity (together with its inferior elements) can be represented by a universal algorithm, which is described by the SEA process model, the goal setting and technology creation model, and the resource pool model (Sects. 5.2, 5.3, 5.4; BPMN diagrams Fig. 5.6 and Fig. 5.13).

"Universal control algorithm" can serve as a formal basis for integrating and integrating various management mechanisms.

Also, from the statement about the composition of the CA and the statement about the multiagent presentation of the CA, a stronger assertion about the *aggregated composition of the CA* follows.

Any set of control elements of activity can be represented by a single control element that implements the compositions of the corresponding components of management: the aggregate of subjects of the initial control elements of activity unites into a single subject, goals, tasks, technologies are combined and harmonized, and so on.

This implies the assertion

Aggregated view of complex activity

Any complex activity (as well as any of its elements that are complex activity) can be represented in the form of an aggregate of elements of the following types organized and united by a common goal-setting:

- specific elementary operations (representing elementary activity), united by a single logical and single causal structure;
- the only control element of the activity that implements analysis, synthesis, specification, regulation and evaluation of CA.

Obviously, the subject of the aggregated control element will be combined will unite all the subjects of the initial elements of the CA, similarly, the goals, technologies and other components of the CA.

The subject of the aggregated element of management activity includes both the subjects of specific elements, and the subject of the aggregated control element, that is, *self-organization* and *self-government* take place in this case.

The assertion about the aggregated representation of the CA allows us to consider any element of the CA as a single aggregated CSA that integrates all the downstream elements of the CA. The logical structure of the aggregated SEA includes all the specific elementary operations of all initial SEAs, the cause-effect structure is formed as a composition of the cause-effect structures of the initial SEAs, and the aggregated process model contains the rules of all the initial process models.

The statement about aggregated representation requires several explanations.

First, the aggregated representation of more or less complex elements of activity is of purely theoretical importance, and in practice the CA is organized and managed in the form of multi-level hierarchical structures. This fact seems intuitively contradictory, since minimizing the number of organizational and control elements (while preserving their functions) reduces costs and thus should increase efficiency. However, the hierarchies of the elements of activity are more effective than single-level aggregated structures, and therefore they are implemented in practice. An analysis of the causes of this fact was made in Sect. 7.4 when discussing the synthesis of optimal technology.

Secondly, the concepts of organization and management traditionally used in management, theory of a company and related disciplines are broader and less strict than those used in this work. In the disciplines mentioned, the technologies of activity as such are not considered, and therefore the creation of new technologies of activity (which is also a CA!) is included by default in organization and management, which, generally speaking, leads to confusion of concepts.

Thirdly, the possibility of the so-called *ideal aggregation* (which does not lead to loss of information about the aggregated object) is extremely attractive from the point

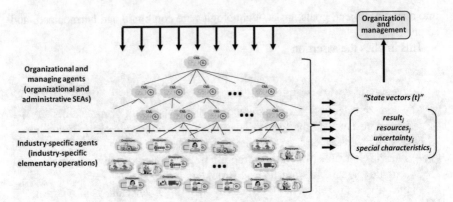

Fig. 7.6 Model of organization and management of CA

of view of a compact and informative description of individual fragments of the CA or of all the activity in general. However, it should be remembered that, having only an aggregated description, it is generally impossible to uniquely *decompose* it, that is, to restore a detailed description, therefore it is necessary to have and store various (and detailed, and aggregated) variants of descriptions using them at different levels of CA analysis.

The statements about the composition of the CA and the multiagent representation of the CA make it possible to describe the organization and management of the CA in the form of a model, which is illustrated by Fig. 7.6.

The element of complex activity under consideration is presented as a set of specific agents (specific elementary operations, particular cases of elements of activity) and organizational and managing agents (organizational and administrative SEAs, elements of activity). The goal-setting, interaction and connections of agents are described by logical, cause-effect and process models of SEA, as suggested in Sect. 5.5. Due to the generation of CA elements, the composition of agents varies with time.

The state of each CA element at each time point is generally described by the characteristics of the results, resources, uncertainty and, possibly, other specific characteristics of the agent (SEA). The controlling agents-SEAs (or the only aggregated such agent) implement the management of this system-depending on the occurrence of certain conditions set on the set of its states, they initiate the execution of certain SEAs, synchronizing the operations and providing the required sequence in time. Organizational (or single aggregated) SEA agents establish links between the subjects of CA elements (generally, subjects of subordinate managers, organizational and specific elements), and also between actors and resources.

The model of organization and management together with the effectiveness metrics introduced in Chap. 6 allow us to formulate and investigate formal problems of optimization of complex activity, which is discussed in Sect. 7.3.

Complex activity is realized in the form of complex hierarchies of its elements, therefore it is necessary to investigate the formation of such hierarchies from the point of view of organization and management. To this end, consider the activity of subjects of several levels: the subject A-SEA, the subject of the superior B-SEA and the subjects of the lower H-SEAs (three horizontal series-level in Fig. 7.7) in all phases of the life cycle of the SEAs (three vertical rows in Fig. 7.7). Chapter 5 revealed that the interaction between SEA is implemented in the form of messaging, so we will analyze how this happens.

The subject of the superior B-SEA or the external environment presents "*demand*" (arrows 1 in Fig. 7.7), which includes:

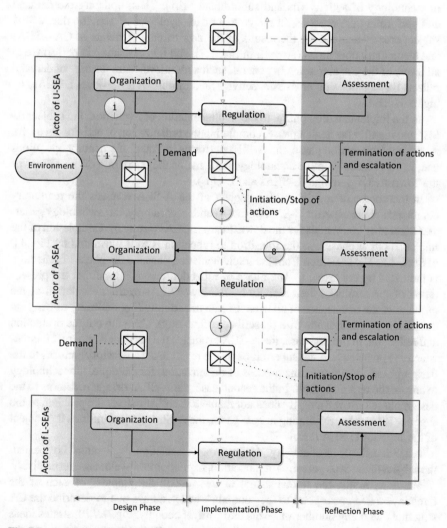

Fig. 7.7 Organization and management in LC of CA

(1) the requirements for the result of A-SEA activity (including the results of subordinate SEA subordinate to it);
(2) requirements and limitations on methods of obtaining the result of A-SEA activity, i.e. on the technology of the latter.

During the design phase of its activity and the activity of lower-level SEA, A-SEA captures demand from the higher SEAs, realizes it and transforms it into its own need, performs goal-setting and structuring of goals/objectives (its and lower SEAs—in relation to them, these will be requirements for their results); formation of technology of activity (its and subordinate SEA). Thus, updating the demand and performing the analysis, in the course of the synthesis of activity, the A-SEA subject creates demand for the results of the downstream elements of CA—SEAs and elementary operations (arrow 2 in Fig. 7.7), and this cascading is performed at all levels of the CA hierarchy. In general, such a multilevel "structuring" of demand reflects the organization of A-SEA activity (and through it lower-level L-SEAs) by higher U-SEA.

To the implementation phase, to the implementation of regulation, the subject A-SEA moves after the design phase, completing the organization of activity (arrow 3). At the implementation phase, the A-SEA performs its "own" elementary operations and, in accordance with the technology, implements the regulation of the actions of the downstream elements (L-SEAs and L-Ops).

In the course of regulation, the subject of the A-SEA receives the regulatory commands of the superior U-SEA (arrow 4) and, accordingly, the technology generates regulating commands for the downstream L-SEAs (arrow 5). Completion of the implementation phase and the transition to reflection is well illustrated by Fig. 4.8 of Sect. 4.5. The transition from the implementation phase to the phase of reflection is indicated by the arrow 6. The phase of reflection comes after either the presentation of the results of their activity by the lower SEAs or their escalation of the problem of the occurrence of an event of true uncertainty (impossibility of carrying out activity in accordance with prescribed technology). Carrying out the evaluation and reflection, in the first case, the A-SEA evaluates the result and either (if it is satisfactory) presents it to the superior, or (if the result is unsatisfactory) returns to the design phase, changing in the process of organization, for example, the technology to ensure the required result. In the second case, the A-SEA, or, again, returns to the design phase (arrow 8), or, if "does not manage itself", escalates the problem to the superior SEA. After evaluating in both cases, the A-SEA subject informs the subject of the superior U-SEA (arrow 7).

After that, the cycle of activity "design—implementation—reflection" is repeated, the subject of the SEA performs organization, regulation and evaluation, respectively.

We note, firstly, the fractal nature of this model: the "inputs" of each of the three layers and correspond to the "outputs", which allows one to describe the CA structures with any number of levels uniformly. Secondly, Fig. 7.7 illustrates ideas as a generalized management cycle of "organization-regulation-evaluation" (arrows 3, 6, 8) and fractality of the hierarchy of SEA (arrows 1, 2, 4, 5, 7).

The statement "On the Composition of Comprehensive Activity" and "On the Aggregate Presentation of the CA" is important to be correlated with the practical implementation of complex activity, namely, to consider the possibility to divide in practice the system-wide, that is, managerial, activity (represented by SEA) and specific activity (modeled by elementary operations).

In practice, almost every CA participant is usually the subject of several elements of activity, combining several system-wide, managerial (including organizational) and specific elements of the CA.

Ordinary executors (for example, workshop workers, bank account managers, ordinary engineers, cashiers and store consultants) usually do not combine or combine in a very small degree the managerial and specific elements of the CA. We can assume that each of them performs one or several elementary operations.

All the other higher-ranking participants in the CA can implement system-wide—management elements of the activity in parallel with specific, or only system-wide ones.

In this case, the "levels" of combining CAs are usually equivalent: it is known, for example, that A. N. Tupolev not only took key design decisions for the aircraft, but also organized project and production cooperation in general. The head of the workshop section solves the minor technological problems of the fitter-picker himself, while the other escalates to the technological department.

In terms of SEA, this means that A. N. Tupolev answered, firstly, for the top-level SEA, the aircraft as a whole, decomposed this SEA into subordinate ones and organized cooperation, and at the same time itself carried out one or several elementary (in terms of this work, i.e., undecomposable) operations determined the total layout, contours, constructive-power scheme, etc.

When managers implement only system-wide management—the elements of the activity, the relevant specific elements of the activity are performed by other subjects (individuals or STS), for example, in the Manhattan Project, General L. Groves performed upper-level organizational and managerial functions, and the corresponding scientific and technical decisions were made by a group of physicists (R. Oppenheimer, E. Fermi and others).

The topic under discussion is closely connected with two trends.

First, the regularization of activity noted in Sect. 4.2 leads to high levels of elaboration and testing of technologies. As a consequence, specific elementary operations are present only at the lower levels of the hierarchy of the CA and are aimed only at obtaining a specific result. At the upper levels of the hierarchy of CA, only management elements that are system-wide remain.

Secondly, the general complication of modern products, products, services—the results of CA—leads, as a rule, to the separation of managerial and specific elements of activity at all levels. For example, if in the middle of the last century a new model of an airplane or a car was designed by a team of 20–50 engineers, now similar cooperation includes hundreds and thousands of specialists, many of whom perform only managerial functions.

7.3 Optimizing Complex Activity

We highlight and consider system-wide aspects of CA optimization, using the statements about the composition of the CA and presenting the CA in the form of a set of interrelated (i.e., organized) and specific elements of management, management and organization that have the same goal-setting. For this, it is advisable to use the formalism of the extended multiagent system, as done in Sect. 5.5 (Fig. 5.18).

Let us consider some (conditionally call it "*target*" and we will mark the CA element with the index i), the SEAs decomposed into a set of subordinate SEAs, conditionally describe them with the set $I = \{1, 2, ..., J\}$. In the general case (if the decomposition includes replicative or creative SEAs) the a priori composition of the set I is unknown, but a posteriori, upon completion of the activity, it is always known.

The *goal*, *structure* and *behavior* of such a system are specified by the logical, causal and process models that make up the CA technology at the system-wide level.

In Sect. 6.3, it was stated that for the CA element its effectiveness and efficiency are completely determined, first, by the content of the "technological predicate", and secondly, by the values the uncertainty characteristics. Efficiency, and the effectiveness of the target element are some known (up to the knowledge of a lot of SEAs) compositions of the performance and effectiveness functions of downstream SEAs, and the "technological predicate" is a composition of the predicates of lower-standing SEAs. CA technology, the "technological predicate", is in general a subject of synthesis at the stage of forming the technology of the life cycle of CA.

The presence of known functions of effectiveness and efficiency theoretically allows us to set a criterion for the effectiveness of the execution of the life cycle of the CA, to put and solve the optimization problem: to achieve the CA goal in an optimal way.

However, this theoretical possibility is not always realized. Depending on the type of uncertainty, completeness and sufficiency of knowledge about the "technological predicate", several variants of posing the optimization problem arise. Based on the classification, given in Table 6.1 of Sect. 6.3, we will define these variants (see Fig. 7.8).

The *task of optimizing the execution of the CA* proves to be correct in practice only for cases of regular activity (Sect. 4.2), when the technology of activity has been created and tested a priori (see Sect. 5.2 for regular activity notes). This option corresponds to the case when the "technological predicate" is completely defined (the left part of the scheme is Fig. 7.8 and the first row of Table 6.1). Characteristics of this option:

- Many lower-level SEAs I and the "technological predicate" are known a priori,
- Substantial uncertainty is measurable and, in the event of an offensive, can only interrupt activity and/or cause recurrence of its elements, without leading to a change in technology.

The optimal performance of such a CA is determined by the choice and assignment of resources and subjects of SEAs and values of characteristics of measurable

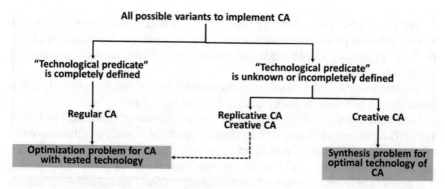

Fig. 7.8 Variants of CA optimization problems

uncertainty. Therefore, the task is solved on the basis of the process model of an element of complex activity (Sect. 5.2) and the resource life cycle model (Sect. 5.4). Measurable uncertainty is eliminated by means of probabilistic approaches. The possibility of an analytical or numerical solution of such a problem is determined by the features of the decomposition of the target SEA, which is a specific factor, so it does not require consideration. A task that is distinguished by these features will be conventionally called the *task of optimizing the performance of a CA with a known technology* and will be considered in more detail in Sect. 7.3.

When the "technological predicate" is unknown or not fully defined (the central and right parts of Fig. 3.4, lines II and III Table 6.1), the set of downstream IEDs and the "technological predicate" are a priori unknown. Therefore, the optimization task in relation to such CA strictly can not even be set.

Nevertheless, we will outline the system-wide characteristics of this problem and formulate constructive approaches to its solution.

Effectiveness and efficiency are determined, as was emphasized above, by the content of the "technological predicate" and the values of the uncertainty characteristics. While the direct performance of the activity, its organization and implementation of management (accurate to the conscientious and trouble-free performance of the elements of the CA, which is a particular case of factors of true uncertainty) only fix the values of effectiveness and efficiency.

Thus, the problem of achieving the highest possible effectiveness and efficiency of complex activity at a qualitative level can be decomposed into two, generally speaking, related tasks:

- a priori synthesis of the optimal technology in the sense of the chosen criterion, the new technology of the "technological predicate" (right-hand side of Fig. 7.8 and the third row of Table 6.1);
- The best operational execution of the CA, taking into account its current state and characteristics of uncertainty and other factors according to the technology, including organization and management (the central part of the diagram and the second row of Table 6.1).

Synthesis of the new technology is a specific and largely heuristic process, which it is not possible to describe with any analytical dependence. Therefore, the choice of "best" technology can be made only on the basis of generation and comparison of several options.

Synthesis of the new technology is the core of the creative CA (Sect. 4.4), it is implemented according to the model described in Sect. 5.3.

Elimination of uncertainty by one of the methods discussed in Sect. 6.3 will allow a priori (before performing the activity) to evaluate the values of the performance and efficiency function. Based on such estimates, the choice of the optimal alternative technologies developed can be made. To assert that the result obtained is optimal, it is possible only after proving a complete comparison of all technologies that are acceptable in some sense.

The problem of synthesis of the optimal technology is considered in Sect. 7.4.

To complete the analysis of the formulation of the optimization problem, it is necessary to consider the remaining version of the problem of the best operational execution of a CA under an incompletely defined "technological predicate". This is an option for optimizing replicative or creative CA realization, when the technology has already been created, the variant corresponds to the second line of Table 6.1 and the central column of the scheme Fig. 7.8.

This option is similar to the task of optimizing the performance of a CA with an approved technology, the difference lies in the presence of true uncertainty. Eliminating true uncertainty with respect to factors that can not be controlled or influenced, one of the methods discussed in Sect. 6.3, will reduce this option to the problem of optimizing CA with an approved technology.

> **Optimization of complex activity**
>
> The task of optimizing complex activity as a whole can be posed and solved in one of two variants: optimizing the performance of CA with known technology and the synthesis of optimal CA technology.

Both variants of setting up the CA optimization problem are considered in the following two sections.

Such optimization problems are extremely widely distributed in practice. On the one hand, they constitute the subject of such a field of applied knowledge, as the section of management, called operation management (see, for example, [45]). They are put in the form of increasing operational efficiency, and such well-known methods and tools as LEAN, TQM, "6 sigma" and others are used to solve them (see, for example, Ref. [60] for a review of such methods). On the other hand, they are the central part of the section of applied mathematics known as operations research (see, for example, [138, 150]).

In our illustrative examples, the tasks of synthesizing optimal technology include the development of:

- the whole set of rules for the actions of employees of the retail bank's branch for servicing clients (operations on loans and deposits, plastic cards, utility bills, etc.);
- technological maps of assembly operations for the assembly of a certain aircraft unit;
- the procedure for firefighting operations at departure to eliminate the emergency situation;
- technological maps of repair operations of a certain unit, system or installation.

The tasks of operational optimization of CA realization with approved technology arise and are solved, for example, when assigning employees for the performance of the abovementioned works of customer service, assembly, repair.

In concluding this section, we note that the problem of aggregation also exists for the values of the efficiency of organization and management. Indeed, how, knowing the effectiveness of the management of lower SEAs, find the effectiveness of management of higher (relative to them)? The above recommendations on measurement scales and monotonicity remain valid for this case. Moreover, the question arises about the possibility of *decomposition and/or integration of solutions to the problems of optimization of organization and management*—after all, using for each of the SEAs the corresponding locally optimal management, one can not be sure that the management of the whole CA is optimal.

7.4   Problems of Optimizing the Execution of Complex Activity with a Known Technology

The tasks of optimizing the realization/execution of complex activity with known and approved technology will be considered on the basis of the analysis of the realization of the life cycle of a CA in time, for the description of which we use the model models developed above, especially the process model (Sect. 5.2). First, let's consider at the system-wide level the process of implementing the A-SEA in time, the corresponding scheme is shown in Fig. 7.9. Then we select optimization tasks that can be solved at each of the implementation phases and for the entire CA as a whole.

At the bottom of Fig. 7.9 denotes the time axis, the phases of the LC of the CA and the times of the beginning and the end of the phases.

The simulated A-SEA is presented in the upper "path", in the middle lane the lower L-SEAs (L-SEA$_1$ and L-SEA$_j$) are shown, and in the lower—the resource pools. Each downstream L-SEA due to fractality is realized in the form of the same scheme, which is reflected in the form of callouts to the right of L-SEA$_1$ and L-SEA$_j$.

The symbols of SEAs and elementary operations are arranged in the figure so that to conditionally show their location on the time axis is that they begin and end at different times.

The mutual arrangement of the intervals for the performance of the elements of activity is determined by the cause-effect model, and the moments of the beginning

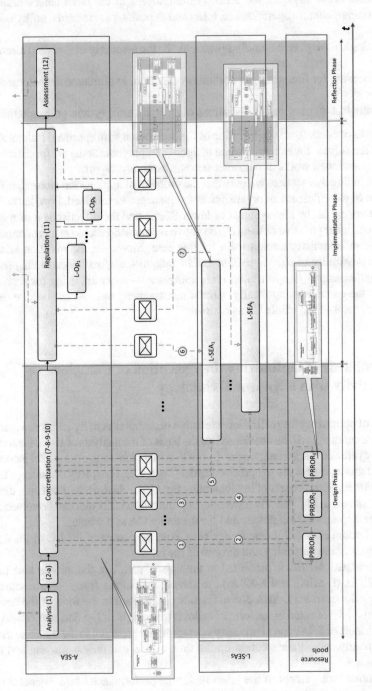

Fig. 7.9 Time schedule of SEA execution

and completion are also the onset of uncertainty events. Since the exact values of the time points are not of interest for the analysis being performed, the influence of the uncertainty factors is not yet considered.

The execution of A-SEA in time begins (Sect. 5.2) from the design phase, from the analysis of demand and other factors (the "Analysis" element in Fig. 7.9 numbers in parentheses correspond to the elements of the BPMN-model elements in Fig. 5.4 and the stage numbers in Table 5.1). After this, the technological information is read (element 2-a) and "Specification (7-8-9-10)" is performed. A-SEA subject concretizes the causal structure in the form of calendar-network and resource plans.

In parallel, the A-SEA entity coordinates, requests and assigns resources that provide L-Op technologies (arrows 1 and 2), and L-SEA subjects (arrows 3 and 4), implementing the procedure for querying, obtaining and organizing resources (Sect. 5.1, Fig. 5.2).

Appeals are performed for all downstream elementary operations and SEAs (rectangles $L\text{-}SEA_1$ and $L\text{-}SEA_j$) entering the logical structure of A-SEA. Obtaining a resource for the organization of the subject of the lower L-SEA signifies its decision to start the activity and initiate the life cycle of activity (arrow 5) described by the lower L-SEA, as shown in Fig. 7.9 with the appropriate notation ($L\text{-}SEA_1$ and $L\text{-}SEA_j$).

During this period of time, the characteristics of the *object* do not change, since no *action* is taken, but changes in the characteristics of resources occur, since the subject of the A-SEA forms the structure of the CA and performs operations to provide the LC of the CA.

The result of the CA is formed in the implementation phase, when the subject A-SEA performs the "Regulation (11)" operation, initiating the execution of the children ($L\text{-}Op_1, L\text{-}Op_k, L\text{-}SEA_1$—arrow 6, $L\text{-}SEA_j$). The conditions under which the initiation of the children are performed are specified by the cause-and-effect model of the A-SEA. At the end of the actions (or completion of all actions according to the technology, or termination of them in the event of occurrence of the corresponding uncertainty events), the lower SEAs informs the A-SEA about this (arrow 7).

The implementation phase of the A-SEA also ends with either the completion of all *actions* according to the technology, or the termination of actions when the corresponding events of uncertainty occur. In both cases, the characteristics of resources also change significantly, since during this interval the basic actions are performed, and in the first case the characteristics of the object take the target values—the objective of the activity is directly achieved.

During the reflexion phase, the characteristics of the object remain constant, the characteristics of the resources change insignificantly, since all lower-level operations and SEAs are completed, the only operation performed is reflection "Evaluation (12)".

Let us now turn to the problem of optimizing the process of executing the CA, presented above, for this purpose we single out its essential factors.

(a) The lower elementary operations of the cause-effect model L-Op$_1$, L-Op$_k$, as well as the elementary operations of the process model (included in the "Analysis", "Specification", "Regulation" and "Evaluation") are specific, therefore their performance functions, the effectiveness and content of the "technological predicate" are considered given and are not subject to optimization.

(b) From the point of view of effectiveness and efficiency, the sequence of operations is fully characterized by the moments of their beginning—the moments of the initiation of operations.

(c) The conditions for initiating transactions (and L-SEAs as "aggregates of operations") are specified by a technology that is known and fixed. After eliminating the uncertainty in the part of factors that can not be controlled or influenced, the conditions can include only the characteristics of the results of all operations (and L-SEAs) and the availability of resources assigned by the technology.

(d) Initiation of operations (and L-SEAs) is performed by the subject of the A-SEA according to the conditions specified by the CA technology.

(e) Resources are used to organize the subjects of L-SEAs and to provide technologies for subordinate elementary operations. In the context of this task (the realization of a CA with an approved technology), the resource requirements in both cases are the same: during the interval of use (the time interval when the resources are directly used for the realization of the CA), they must maintain their characteristics in a range sufficient to execute the CA.

Taking into account these factors, the effectiveness and efficiency of CA realization with known technology is determined by the resources:

- Availability of a sufficient number of heterogeneous resources with the required characteristics during the intervals of use;
- The adequacy of the characteristics of each resource during use intervals to ensure the performance of the CA element.

In the sections on resources (Sect. 2.9 and 5.4), it was noted that resources need to be considered not only during use intervals, but also throughout their life cycle. In Sect. 5.4, a life cycle model of the resource pool is proposed, which should be used for setting and solving optimization problems.

In fact, the timing of the execution of the SEA (in fact, a set of such schemes for interrelated SEAs) sets an integrated need for heterogeneous resources.

In response to such a need, the operational structure of the elements of activity is formed—the concretization of the logical and causal models with reference to the calendar time and the corresponding hierarchical network graph (its projection on the objects). The consistency of the timelines (key points of the network schedule) is checked with the need for the results of the activity, with inconsistency, the operational structure is re-formed.

The temporal consistency of the operational structure and the resource pool is checked taking into account the possible changes in the states of non-expendable resources in the process of activity. If there is inconsistency, the operational structure is adjusted. The result is a graph of the dynamics of resource use.

Optimization of the dynamics of resource use is optimized (taking into account the possibility of implementing with the use of these resources in parallel with other elements of the CA). The result is an optimal operational structure.

> **Optimal realization of complex activity with known technology**
>
> The tasks of optimal realization of a CA with an approved technology can be formulated as interrelated tasks:
> - optimal assignment of dissimilar resources taking into account multiple consumer SEA;
> - optimal maintenance in the required ranges of resource characteristics throughout their CC, primarily during use intervals;
> - optimal maintenance of the pool of resources in the required ranges.

7.5 Problems of Designing the Optimal Technology

From the point of view of optimizing the CA, the problem is posed as a synthesis of the "technological predicate", the best in the sense of the chosen criterion, built on the basis of effectiveness and efficiency.

Proceeding from the statement about the composition of the CA and the statement about the aggregated representation of the CA (Sect. 7.1), one can say that the synthesis of the new technology consists in the synthesis of the logical and causal structures of CAs, the creation of technologies for specific elementary operations and technologies of organizational and control elementary operations or SEAs.

The development of technologies for specific operations is not the subject of this work, therefore, within the framework of this task, they should be considered to be set up to the possibility of having several, also specified, variants of technology sets.

According to the allegations about the causes of the cause-effect relations of the CA (Sect. 3.4) and the foundations of the logical structure of complex activity (Sect. 3.3), the specificity of the CA also defines the variants of logical and cause-effect structures due to the presence of natural specific links between elementary operations. Obviously, for example, that the final assembly operations of an aircraft can not precede assembly operations or workpiece processing operations for the manufacture of a particular part.

It is appropriate to consider the reasons for the emergence of hierarchies of complex activity. This question is closely connected with the problems of determining the optimal organizational structure of a firm, enterprise, organization, division (as STS), widely discussed in certain sections of management, and some authors consider the organizational structure a key factor in the firm's effectiveness (see, for example, the classic work of Mintsberg [91], and also [9]).

As noted above (see the statement about the aggregate composition of the CA in Sect. 7.1), the realization of the CA in the form of multi-level hierarchies seems irrational: management and organization, according to this statement, can be carried out by a single element without loss of functionality, and the presence of many such elements increases costs. Let us analyze why, nevertheless, complex activity is realized in the form of hierarchies, and not aggregated SEA.

The main objective reasons for the existence of hierarchies of CA are connected, first of all, with the possibility of manifesting true uncertainty in any CA. From this possibility it follows that "any significant" SEAs can not consist only of regular elements, but will also contain creative elements—that is, contain activity to create new CA technologies (in order to ensure the response to a priori unknown events of a true uncertainties). In practice, "somehow complex" activity, "any significant" element of the CA, can not be completely regular: aircraft design, the work of a retail bank, a fire department or nuclear power plants during the reporting period (month, quarter, year) require at least planning and rescheduling—that is, creating and/or modifying the technologies of CA elements.

Suppose that some "any significant" A-SEA is implemented in the form of an aggregated view: it includes a significant number of downstream specific elementary operations and no downstream SEA. That is, this aggregated A-SEA implements the organization and management of a significant number of elementary operations without an intermediate hierarchy of organization and management. This means that the subject of the A-SEA should react to everything and any uncertainty events that arise during the performance of elementary operations. It is also obvious that an unknown event of uncertainty a priori can be heterogeneous and, therefore, require different reactions—different resources, different competences and responsibility in making decisions. However, the absence of intermediate SEAs leads to the fact that the subject A-SEA is forced to react equally to all such heterogeneous events. And this, in turn, means excessive spending of resources, competencies and responsibilities.

Thus, the aggregated system does not allow differentiating the resources, therefore it is not effective.

Hierarchical structures, in turn, ensure the possibility of adequate and differentiated delegation of responsibility to lower levels, network structures of risk sharing; therefore such structures are used in practice.

Synthesis of a new technology (or modernization of an existing one) includes the following steps (Sect. 5.3):

(a) the design of the logical structure of the CA;
(b) the creation of technology downstream in the logical structure of SEAs and elementary operations;
(c) designing the causal structure;
(d) creation of a process model;
(e) verification of the feasibility of the technology.

Consider these steps, define the parameters that allow controlled (for optimization) changes and select the optimization tasks to be performed on each of the steps.

(a) Design of the logical structure of the CA. The first step is the formation of a logical structure—a hierarchy of goals and sub-goals, starting with the main goal of the highest level. Direct formation of the hierarchy of goals does not generate optimization problems, because, firstly, the logical structure is specified by the specifics of the technology, and secondly, the hierarchy "in itself" does not cause the consumption of resources and, therefore, is not characterized by performance indicators.

(b) Creation of downstream technologies in the logical structure of SEA and elementary operations. Each of the goals requires, for its achievement, the realization of an associated CA element, which means that the entire set of constituent elements of the SEA is associated with each of the objectives. The result and the subject (through the result) are already aligned with the goal structure in the previous step. Actions are a reflection of technology and therefore are not considered in the synthesis of technology. Therefore, at this step, the goals need to align the subject and technology. Subjects and technologies are formed on the basis of resources, therefore, as a result, the requirements to the structure of the resource pool are defined, these requirements include specification of the characteristics of resources and their scope. The structure of resource pools can and should be the object of several optimization tasks, and, in general, iteratively repeating ones.

(c) and (d) Design of the cause-effect structure and creation of the process model. The cause-effect structure itself is determined by specific factors, and optimization is not possible. When these steps are performed, the cause-effect and time relationships between goals/results are fixed, which determines the resource requirements taking into account the moments and time intervals when the resources should be used. At this stage, the requirements for resource pools are specified, concretizing the setting of optimization tasks.

(e) Verification of the feasibility of the technology. At this step, optimization tasks are not generated.

Thus, the creation of new technologies generates several optimization tasks related to resources:

- optimization of the set of resource pools based on the needs of the technologies of various elements of the CA;
- optimization of the volumes of each resource pool;
- optimizing the maintenance of resource characteristics within specified ranges during their entire LC.

Synthesis of new technology generates requirements for resource pools, including specifications of resource characteristics and their volume at various times and intervals.

Following the requirements, it is necessary to check the existence of pools of necessary resources (by comparing the characteristics of existing and necessary ones) and the sufficiency of their volumes.

Depending on the results of the audit, new elements of activity for implementing the life cycle of resources can be organized or modernization of existing resource pools can be carried out (the execution model of the life cycle of resources is described in Sect. 5.4).

The initial data of the tasks are the forecast of resource requirements, models of own resource behavior and models of maintaining resource characteristics in the required ranges. The forecast of the need and model is formed taking into account the elimination of uncertainty in the ways considered in Sect. 6.3.

Because of the possibility of forming several variants of specific technologies in the creation of new technologies, the problem arises of choosing the optimal variant, which is solved in the same ways as the tasks related to optimization of resources.

The performed analysis of the process of creating a new technology allows us to formulate the following statement.

Synthesis of optimal technology

The synthesis of technology is essentially specific, and therefore itself allows optimization only to the formation of several alternative options and choosing the best among them.

Synthesis of technology at the same time generates resource requirements and related optimization tasks:

- optimization of the set of resource pools based on the needs of the technologies of various elements of the CA;
- optimization of the volumes of each resource pool;
- optimizing the maintenance of resource characteristics within specified ranges during their entire LC.

The creation of new technologies includes a system-wide part (generally speaking, a secondary one in this case), for which the optimization tasks related to resources are listed above, as well as a specific part—the synthesis of elementary operation technologies, logical and causal structures. Due to the specificity of this part, it is impossible to constructively formulate the corresponding optimization tasks at the system-wide level. However, for it a number of system-wide recommendations or requirements can be given, which in turn will provide the basis for the synthesis of alternative technologies and thus provide the choice of the best option. Let's formulate system-wide recommendations and requirements.

In the previous sections (introduction, 2.7, 4.2, 4.4, 5.3), the significant role of uncertainty factors has been repeatedly noted, which complicates the realization of complex activity in general and the creation of new technologies in particular.

In the process of creating new technologies, the uncertainty of one of the main groups is manifested and has the greatest impact (see the grouping in Sect. 2.7)—the uncertainty of technology.

The developed system of CA models allows to naturally divide elements of CA according to the degree of "complexity and uncertainty" and appropriately organize resources, sending them "there", where for overcoming uncertainty, much more costs (management and material resources, time costs) are required in comparison with routine (less complex and indefinite) elements. This opportunity is a generalization of the widespread practical approach of TRL—Technology Readiness Level (see Sect. 4.2 and [77, 141], which assumes the definition of formal levels of technology maturity and the application in practice of only high availability technologies.

F. Knight in Chapter 8 of [70] recommended to overcome the uncertainty:

- reduce the levels of uncertainty by grouping events of uncertainty and thereby estimating probabilities of outcomes based on the accumulation of statistical data;
- limit the number of persons who are called upon to make decisions in conditions of uncertainty and, therefore, bear responsibility for it;
- "manage the future" by predicting possible options for its occurrence;
- "distribute" the consequences of uncertainty to several entities;
- avoid activity associated with uncertainty.

Based on TRL principles, F.Knight approaches and the CA models presented in the previous chapters, we formulate a number of system-wide recommendations for the creation of new technologies. Again, we consider the creation of a technology for the operation of a certain SEA, which includes specific technologies for downstream elementary operations (L-Ops). The development of technologies for downstream SEAs is of a system-wide nature and therefore is of no interest in this case.

First, all technologies can be divided into groups on the basis of maturity, elaboration, "approbation", readiness. To the elements of CA, the technologies of which belong to different groups, it is advisable to apply different sets of instruments of organization and management. In the most general case, there can be three such groups. The first and third of them reflect polar cases, and the second—obviously existing, intermediate.

Group 1. Use of known technologies without changes or with changes that obviously do not affect technological uncertainty. For example, replicating the manufacture of a known part (or unit or end product) using a known technology on a new production site, or launching the manufacture of a known part using known technology, but with minimal dimensional changes. These are cases when the technology of a new element of activity is a priori known and actually copied.

Group 3. Fundamentally new technologies with a significant level of true technological uncertainty, for which an assessment of the possibility of a successful completion can only be carried out on the basis of subjective judgments of experts. For example, the use of new principles for the processing of materials for the manufacture of parts or the performance of repair operations in conditions where the essential characteristics of the repair object, affecting the labor and the result of repair, are unknown a priori and can only be assessed during the repair work itself. In fact, this is the case when the technology of activity is created in the course of performing activity.

Group 2. Technologies that do not belong to either group 1 or group 3 cover all intermediate cases in which a significant share of technology is known a priori. *Technological uncertainty* can be either measurable, or true, but not having a significant effect on the final result.

Secondly, we will define system-wide grounds, based on which we formulate recommendations. In Sect. 6.3, when considering approaches to eliminating uncertainty, it was found that in the case of true uncertainty with completely unknown a priori characteristics of events (case III, third row of Table 7.2), all or some of the components of the activity and/or the external environment have not been adequately studied. Therefore, the performance of activity is in this case the acquisition of new knowledge, which leads to a change in technology in the implementation process.

Therefore, the natural basis for the recommendations will be a more accurate and detailed fixation of the elements of activity with a low level of uncertainty, which will allow us to more accurately specify the "domain of definiteness", the knowledge already available, and focus on really vague elements. Therefore, system-wide recommendations should concern a more precise, detailed description of the elements of the CA on the subject, actions, technologies and other grounds and the localization of the uncertainty or expansion of the "domain of definiteness".

Thus, we can formulate the following recommendations for the organization of the process of developing specific technologies.

(a) With sufficient detail to structure the object of the CA, in order to more accurately localize the elements of the object, which are highly uncertain and at the same time to identify standard, typical elements that allow the use of ready-made solutions. For example, one of the main trends of the current stage of development of production technologies is the widest possible use of purchased components; in particular, about 65% of the costs of all components and components of a modern car is the costs of standard components.

(b) With a sufficient degree of detail to structure CA technologies (directly related to recommendation a)—to separate the elements of activity related to the various groups 1-2-3 above and differentially apply different approaches for them, which increases the efficiency of performance of activity. In practice, this recommendation is implemented with respect to both specification of the specifications of technological operations and planning.

(c) To develop and consider alternative solutions for both the technology of activity and the subject and its elements. The development of several alternative solutions in conditions of high uncertainty is equivalent to the formation and verification of several hypotheses regarding the a priori of an unknown object, which increases the amount of knowledge concerning it.

(d) Use a scenario approach in forecasting the future. Forming scenarios of possible development of events also increases the amount of knowledge (albeit subjective) regarding an uncertain future and allows using the grounds for creating technologies. For example, the development of a development plan for a retail bank (a special case of development of CA technology) is almost always based on certain long-term forecasts—scenarios for the movement of financial and stock markets and other markets, despite all the subjectivity of such forecasts.

(e) Provide for interim and preliminary actions to integrate the elements of the object and technology elements into a single system. The uncertainty of any complex system is directly related to the complexity and emergence (Sect. 2.7), which often leads to significant problems in the course of system integration. For example, the creation of a new model of a modern aircraft or a car involves the development of a large number of heterogeneous systems, units and nodes. Due to the different nature of the components of the product, their development is carried out by separate groups of specialists, often working in different firms and sometimes in countries. Therefore, the integration of the final product in this case is a complex process, consisting of several intermediate steps.

(f) Provide additional actions to clarify the need—more precise specification of the required characteristics of the subject (Sect. 6.1). Complexity, emergence and uncertainty of the object and goal of CA as a complex system almost always generate incompleteness of knowledge about the need in the early stages of CA. That is, absolutely natural is the situation in which not only the subject, but also external consumers of the CA result, inexactly estimate all the target values of the CA object characteristics. Therefore, in recent decades, in addition to the traditional and obvious ways of clarifying needs through communication with consumers, special management tools have also been widely disseminated in the face of uncertain needs, such as Agile and SCRUM [117, 134].

(g) Provide additional rules for intermediate control of activity—the evolution of the characteristics of the subject (Sect. 6.1) for the earliest possible detection of deviations in the actual dynamics of changes in the characteristics of the object from the required. In the modern practice of project management, design programs, life cycles of products, the principles of setting and performing checks, called checkpoints, control lines, gate of decision-making are widely applied. Each of these checks is a priori specified by several groups of rules. First, these are the rules that determine the moment of verification, they can be set both on the basis of the execution time (after so many days after the commencement of work), the completion of the stages or stages of action, the degree of change in the characteristics of the object. The second group of rules specifies the totality of the characteristics of the object and their combinations that are subject to verification. The third group—determine the necessary follow-up actions depending on the results of the audit.

(h) To provide for a greater degree of control over CA technology, rather than its result. Such control allows, first, to detect problems as early as possible, and secondly, to reveal not only the manifestation of problems in the form of deviations from the actual characteristics of the object from the required, but also the causes of their occurrence, lying in the technology. This approach has also become widely used in practice in recent decades: various quality standards have become popular, for example, the ISO9000 standards system, and the maturity models of production processes (CMMI and other similar tools).

(i) Provide additional control over the performance of activity by subordinate entities. This recommendation, related to the previous one, also provides for earlier identification of problems: before the result of the CA of the subordinate subject is presented to the consumer. For such monitoring, not only control of intermediate results, but, first of all, CA technologies of downstream subjects is also used. In practice, large companies and government agencies often require suppliers (subordinate entities) to pass independent certification for quality standards.

The application of the above recommendations provides more accurate estimates of the complexity and, consequently, the resource intensity of the elements of the activity due to the decomposition of activity and the subject into more detailed elements. As a result, it becomes possible to reduce the resource intensity of activity due to the differentiated optimization of the resources required by the elements of the activity of various uncertainties, including the expansion of the number of regular elements of activity, that is, proven technologies and proven performance results. Additional checks and intermediate integration allows you to detect problems at earlier stages, therefore, to avoid losing the continuation of inappropriate actions, before proceeding to an alternative way of implementing the activity.

Naturally, the use of any of these recommendations entails additional costs, since it requires additional elements of organizational and managerial activity. Therefore, they should be used to create fundamentally new technologies, the third group, and to a much lesser extent—using known technologies (group 1).

Summary of This Chapter
The role and place of organization and management in the system of elements and the life cycle of the CA (Sect. 7.1) are considered. It is shown that any complex activity includes elements related to both organization and management. The structure of the organization and management, as well as the main activity of the subject at various stages of the LC CA (Sect. 7.2), are described.

The statements on the composition of the complex activity and on the aggregated presentation of the CA are formulated and substantiated. The assertion about the multiagent representation of CA proposed in Sect. 5.4 is substantiated. These statements make it possible to describe any complex activity as an aggregate of elements (see Fig. 7.6) organized and united by a common goal-setting, logical and causal structures, of the following types:

- specific elementary operations (representing elementary activity);
- the only or several control elements of the activity that implement (constant, or multiple, or continuous) verification of the occurrence of certain conditions and initiation of the relevant specific elements of activity, and also establish links between the CA entity as a whole, resources and subjects of the subordinate specific elements.

Such a description and model of organization and management of complex activity (see Fig. 7.6) allows:

First, formally divide the elements of the activity into specific (that is, directly form the concrete final result) and managers, which, in turn, lays the foundation for further development of MCA in the direction of the study of such an object as "a system for managing complex activity and STS as a subject of this activity".

Secondly, to formulate the problem of optimization of complex activity and justify the statement (Sect. 7.2) on optimization of complex activity, which defines two areas of optimization: optimizing the performance of CA with known technology and the synthesis of optimal CA technology.

Consideration of the task of optimizing the performance of CA with the known technology is performed (Sect. 7.3) on the basis of the time schedule of the SEA execution (see Fig. 7.8). The resulting statement "On the optimal implementation of CA and known technology" characterizes the corresponding optimization tasks of the organization and use of resources:

- optimal assignment of dissimilar resources taking into account multiple consumer SEA;
- optimal maintenance in the required ranges of resource characteristics throughout their LC, primarily during use intervals;
- optimal maintenance of the pool of resources in the required ranges.

Analysis of the problem of synthesis of optimal CA technology (Sect. 7.4) allowed to formulate and substantiate the statement "On the synthesis of optimal technology". The synthesis of technology is essentially specific, and therefore itself allows optimization only to the formation of several alternative options and choosing the best among them.

Synthesis of technology at the same time generates resource requirements and related optimization tasks:

- optimization of the set of resource pools based on the needs of the technologies of various elements of the CA;
- optimization of the volumes of each resource pool;
- optimizing the maintenance of resource characteristics within specified ranges during their entire LC.

The core of the creation of new technologies is a specific synthesis of the technologies of elementary operations, logical and cause-and-effect structures. Due to the specific nature of this part, it is impossible to put optimization tasks on it at the system-wide level, but for it a number of system-wide recommendations or requirements are formulated, which in turn create the basis for the synthesis of alternative technologies and provide the choice of the best option. Concluding the seventh chapter, we emphasize that practically all the wealth of the modern apparatus of applied mathematics can be used to model and optimize the phenomena and processes inherent in different phases, stages and stages of CA; some of its sections, as well as related areas, are mentioned in Table 7.3.

Table 7.3 Some models and methods used in different steps of CA's LC

Phase	Stage	No.	Step	Models and methods
Design	I. Fixing demand and understanding needs	1	Fixing demand and understanding needs	• Strategic planning • Analysis (in particular, scenario analysis) and forecasting of environment • Data analysis • Marketing • Queueing theory
	II. Setting goals, structuring goals and tasks	2	Creating logical model	• Classification • Integrated assessment • Decomposition/aggregation • Expertise technologies • Multicriteria decision-making • Financial analysis • Optimization (if actors, results and technologies can be chosen) • Mathematical logic
	III. Selecting and developing technology	3	Checking the readiness of technology and the sufficiency of resources	• Resource allocation methods • Mathematical logic • System optimization (if resources can be chosen)

(continued)

Table 7.3 (continued)

Phase	Stage	No.	Step	Models and methods
		4	Creating cause–effect model	• Graph theory • Finite automata • Markov chains • Differential equations • Logistics • Integrated assessment • Data analysis • Scenario analysis • Reliability theory • Probability theory • Fuzzy sets • Interval analysis • Decision theory
		5	Creating technology of lower-level elements	See steps 2–4
		6	Forming/modernizing resources	• Calendar-network scheduling and control • Decomposition/aggregation • Modeling of business processes • Discrete optimization • Scheduling theory • Queueing theory • Game theory • Decision theory
		7	Calendar-network scheduling and resource planning	See steps 4 and 6
		8	Performing optimization	• Continuous and discrete optimization • Multicriteria decision-making

(continued)

Table 7.3 (continued)

Phase	Stage	No.	Step	Models and methods
		9	Assigning actors and defining responsibilities	• Decomposition/aggregation • Modeling of business processes
		10	Allocating resources	• Dicsrete optimization • Game theory • Queueing theory
Implementation	IV. Performing actions and obtaining results	11	Performing actions and obtaining results	
Reflection	V. Assessing results and reflecting	12	Assessing results and reflecting	• Strategic planning • Financial analysis • Decomposition/aggregation • Integrated assessment • Data analysis • Expertise technologies • Decision theory

Conclusion

Concluding this book, we present an *ontology of the main results* obtained above in form of a diagram (see Fig. A.1), and also a *comparative analysis* of general methodology (GM, the methodology of elementary activity) and the methodology of complex activity in tabulated form (see Table A.1).

M. V. Belov and D. A. Novikov, *Methodology of Complex Activity*, Studies in Systems,
Decision and Control 300, https://doi.org/10.1007/978-3-030-48610-5

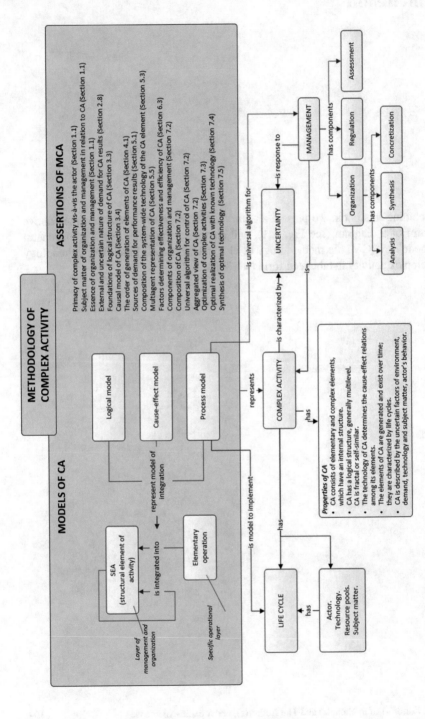

Fig. A.1 Ontology of main results of MCAPlease, change by high-quality picture (file is attached to the letter)

Table A.1 Comparative characteristics of MCA and GM

Characteristics	GM-based knowledge	Methodological knowledge specific to CA	
		Elementary activity	Complex activity
Characteristics of activity			
Distinctions of activity	GM-based descriptions for distinctions of activity organization	1. Routine and predictable character; limited effort by actor 2. Limited adaptivity within fixed technologies; no capability for goal-setting, as the goal of activity is formulated beforehand and from outside 3. Limited opposition to destructive tendencies, limited self-organization and self-development within fixed technologies 4. Invariable staff and structure of activity 5. Invariable technology, actor, and role of subject matter in its goal context 6. Unique actor, subject matter, and technology of activity	1. Unique and unpredictable character; free will of actor 2. Adaptivity; capability for goal-setting, the goals of productive activity are formulated during the activity itself; 3. Strong opposition to destructive tendencies; self-organization, self-development; 4. Variable staff and/or structure of activity 5. Variable technology, actor, and role of subject matter in its goal context 6. Multiple actor, subject matter, and technology of activity
Principles	Same for different types of activity	Specific for different types of activity	1. The principle of correspondence; 2. The principle of complementarity; 3. The principle of completeness; 4. The principle of compactness; 5. The principle of generalization and abstractness

(continued)

Table A.1 (continued)

Characteristics	GM-based knowledge	Methodological knowledge specific to CA	
		Elementary activity	Complex activity
Conditions of activity		Motivational, personnel-related, material and technical, scientific and methodological, organizational, financial, normative and legal, informational conditions of activity	Life cycles of resources, knowledge and organizations
Uncertainty		As a rule, measurable (except for research and art activity)	True and/or measurable Uncertainty of demand, needs, goals and tasks, conditions, requirements and norms, actions and result, assessment
Optimization		Choice of efficient forms, methods and means of activity	Optimization of organization and management of CA
Norms of activity	GM–norms		
Logical structure of activity			
Demand	Not considered	Not necessary	The procedure of demand generating (PDG)
Needs	GM-based definitions of needs, actor, subject matter, and result of activity		Has its own life cycle
Actor		Unique actor, fixed-structure OTS (possibly, of variable size)	Multiple actors in form of OTS with time-varying staff, structure and properties
Subject matter, result		Unique subject matter and unique result	Multiple subject matter and/or multiple result
Technology	Specific or even not defined within GM	Unique and invariable technology	Multiple and/or variable technology
Structuring of activity		Not necessary	Necessary
Typical element of structure		Sequence of procedural components of activity (Fig. 1.2)	SEA (Fig. 3.2)

(continued)

Table A.1 (continued)

Characteristics	GM-based knowledge	Methodological knowledge specific to CA	
		Elementary activity	Complex activity
Classification bases for typical elements of structure		Not necessary	By the level of hierarchy: elementary (indivisible)/complex; By subject matter: thing/knowledge/people (organizational system)/system; By the capability to generate new activity: regular/replicative/creative; By the effort of actor: binary choice/binary choice + uncertainty elimination; By the sources of uncertainty: measurable pool/measurable pool + needs/measurable pool + technology
Formation bases for logical structures		Not necessary	By structure (hierarchies/fractalities): subject matter/result; technology/actions; actor (organizational structure).
Organizational forms, methods and means of activity	GM-based definitions of organizational forms, methods and means of activity		
Temporal structure of activity			
Phases of activity: I. Design phase; II. Implementation phase; III. Reflection phase.	GM-based definitions of organizational phases, stages and steps of activity in the project paradigm		
	Stages of elementary activity: (1) Conceptual; (2) Modeling; (3) Design; (4) Technological preparation; (5) Implementation of models; (6) Presentation of results; (7) Assessment and reflection		Stages of CA: (1) Fixing demand and understanding needs; (2) Setting goals and structuring goals and tasks; (3) Selecting and developing technology; (4) Performing actions and obtaining results; (5) Assessing results and reflecting

(continued)

Table A.1 (continued)

Characteristics	GM-based knowledge	Methodological knowledge specific to CA	
		Elementary activity	Complex activity
Organizational forms	(1) Elementary operations/jobs/processes; (2) Projects		(1) Elementary operations; (2) Complex operations; (3) Projects; (4) Programs; (5) Life cycles
Life cycles as organizational forms of activity	Life cycle of activity		Coordinated life cycles of demand, CA, its subject matters, actors, knowledge, technologies, and resources
Temporal structure of CA in form of SEAs	Not defined within GM	Trivial for elementary activity	Formation bases for the hierarchies of SEAs in time: by subject matter and result; by technology and actions; by actor (organizational structure)
Generation model of new elements of activity			Implementation model of CA in time: permanent SEAs/promptly generated SEAs; regular/replicative/creative SEAs
Process model			Implementation model for the life cycle of CA, the process model of an element of CA
Organization and management	GM as the science of activity organization		Management, consisting of organization (analysis, synthesis, and concretization), regulation and assessment

Appendix A
Basic Notation and Abbreviations

CA—complex activity;

GM—general methodology;

IM—informational model;

LC—life cycle;

L-SEA—lower SEA;

MCA—the methodology of complex activity;

NPP—nuclear power plant;

Op—elementary operation;

OTS—organizational and technical system;

PDG—procedure of demand generating;

PRROR—procedure for requesting, receiving and organizing resources;

SEA—structural element of activity;

SoS—system of systems;

U-SEA—upper SEA.

M. V. Belov and D. A. Novikov, *Methodology of Complex Activity*, Studies in Systems,
Decision and Control 300, https://doi.org/10.1007/978-3-030-48610-5

Appendix B
Basic Definitions

Term	Definition
Activity	A goal-oriented active effort of an individual
Life cycle	The evolution process of a system, product, service or another object, from the origin of concept (or appearance) to disposal (or extinction)
Measurable uncertainty	A possible occurrence of a priori unpredictable events that happened earlier and/or satisfy fundamental laws
Information model	A model of an object with information that describes its essential parameters and variables, the existing connections among them as well as the inputs and outputs of this object, in order to simulate its possible states in response to any variations of the inputs
True uncertainty	A possible occurrence of unique (or infrequent) events that do not satisfy the existing fundamental laws and have no a priori observations
Complex activity	An activity with a nontrivial internal structure, multiple and/or varying actor, technology and the role of subject matter in its goal context
Methodology	The science of activity organization
Uncertainty of CA	A possible occurrence of a priori unpredictable events in the course of CA that affect its implementation and result
Organizational and technical system	A complex system that consists of people, technical and environmental elements

(continued)

M. V. Belov and D. A. Novikov, *Methodology of Complex Activity*, Studies in Systems,
Decision and Control 300, https://doi.org/10.1007/978-3-030-48610-5

(continued)

Term	Definition
Organization	A complex activity to obtain an internal arrangement and coordinated interaction of more or less differentiated autonomous elements of its subject matter (particularly, by establishing and maintaining relationships with given characteristics among these elements)
Openness (of a system)	Free and unlimited (by artificial factors) participation and interaction of elements with each other and the environment
Behavior (of an individual or actor)	A sequence of his/her actions, an interaction with the environment, mediated by his/her external (motor) or internal (mental) actions
Behavior (of a system or object)	A successive (in time), at least partially observable, responsive, measurable objective fixation of state changes
Need/demand	A requirement to the result of activity (the goal characteristics of its subject matter) and ways to achieve it (technologies), including external norms and assessment criteria, conditions and time limits
Subject matter of research	An aspect of an object of research that is studied in a particular case
Subject matter of activity	A specific object that is changed during activity
Extended enterprise	A system of autonomous interacting organizational and technical systems that share the same goals and technology of operation
Effectiveness	A degree of compliance between the result and goals of activity
Resource	Every thing used for a particular goal, including the goal-oriented activity of an individual or a group and activity itself; a quantitatively measurable possibility of performing a certain activity by an individual or a group; conditions under which a desired result can be obtained by certain transformations
System	An assemblage of elements that are related to and connected with each other and form a distinct totality, a unity
System (artificial)	A set of interacting elements, organized to achieve one or more declared goals
Systems engineering	An interdisciplinary approach that directs and coordinates all the technical and managerial efforts required to create a system and implement it in the face of a plurality of stakeholder needs, expectations, and constraints throughout the life cycle of the system
Actor	A carrier of a practical activity, a source of active efforts directed to its subject matter

(continued)

(continued)

Term	Definition
Complex system	A system possessing the property of emergence; an open system with continuously interacting and competing elements
Structure	A set of stable connections among elements of a system
Technique	A set of artificial means of activity
Technology	A system of conditions, criteria, forms, methods, and means successively achieving the defined goal
Management	A complex activity ensuring the impact made by a manager (the actor of this CA) on a system being managed (the object of management) designed to ensure that its behavior leads to the achievement of the entity's objectives
Product's life cycle management	An application of a consistent set of tools intended to jointly create, control, disseminate and use information about a product within an extended enterprise, from the original concept of the product to the end of its life cycle, with the integration of employees, industrial processes, production systems and information
Elementary activity	An activity without any nontrivial internal structure of goals, technologies and result
Emergence	A property of systems stating that the properties of the whole are not reducible to the totality of the properties of the parts from which it is made up and are not derived from them
Efficiency	The ratio of a result obtained to the characteristics of resources consumed

Appendix C
Basic Assertions

Assertion	Formulation
Primacy of complex activity vis-à-vis the actor (Sect. 1.1)	Activity in the form of intention/need arises no later than its actor does as a single individual, an elementary STS. At the same time, a complex activity is primary in relation to its actor, a non-elementary STS
Subject matter of organization and management in relation to CA (Sect. 1.1)	In managing and organizing a CA and/or STS, the subject matter is complex and includes a complex interacting and interrelated pair: {CA ⇔ STS} Complex activity is the primary subject matter of organization and management, and the sociotechnical system itself is an intermediary, playing the roles of the actor involved in the CA and/or its subject matter
Essence of organization and management (Sect. 1.1)	Both management and organization (as a process) are types of activity, usually complex ones, the subject matter of which is the CA and the corresponding STS. At the same time, the indirectness and obliqueness of the influence on the final result through the subject matter of management activity is a distinctive feature of "management," and the goal in the form of a change in internal order and coordination of interacting parts is a distinctive feature of "organization."
External and uncertain nature of demand for CA results (Sect. 2.8)	Activity (its elements), as a rule, is generated as a response to external factors or events. These factors and events lead the subject to activity and are often a priori uncertain. Since activity is generated by external sources, the users of the result of activity are also external persons (in relation to the CA and its subject). At the moment of generating activity, the individual actualizes external demand, turning it into his inner need, and at the same time becomes an elementary subject of CA

(continued)

© The Editor(s) (if applicable) and The Author(s), under exclusive license to Springer Nature Switzerland AG 2020
M. V. Belov and D. A. Novikov, *Methodology of Complex Activity*, Studies in Systems, Decision and Control 300, https://doi.org/10.1007/978-3-030-48610-5

(continued)

Assertion	Formulation
Foundations of logical structure of CA (Sect. 3.3)	The only constructive basis for structuring complex activity—the selection of the hierarchy of its elements—is the structure / hierarchy of the objectives of the activity The logical structure also reflects the "managerial" hierarchy of subordination and responsibility of the subjects of the elements of the CA for the results—for achieving the goals. The subjects of higher SEAs, together with the responsibility for achieving their "own" goals, are also responsible for achieving lower goals, and, as a consequence, for the activity of lower-level SEAs
Causal model of CA (Sect. 3.3)	The causal model of complex activity reflects the specific factors of its technology
The order of generation of elements of CA (Sect. 4.1)	Generation of elements of complex activity is carried out in sequence only from the higher (according to the logical structure) to the lower ones, including in the case of self-organization
Sources of demand for performance results (Sect. 5.1)	Sources of demand for the results of a new element of complex activity can be only the subject of the superior element of the CA or environmental factors
Composition of the system-wide technology of the CA element (Sect. 5.3	The technology of any CA element is described by the structural, cause-effect and process models together with models of subordinate SEAs and elementary operations
Multiagent representation of CA (Sect. 5.5	Any complex activity can be represented as an extended multi-agent model
Factors determining effectiveness and efficiency of CA (Sect. 6.3)	Effectiveness and efficiency are completely determined, first, by the content of the "technological predicate," and secondly, by the values of the characteristics of uncertainty
Components of organization and management (Sect. 7.2)	Management components are organization, regulation and assessment. The components of organization are analysis, synthesis and concretization

(continued)

(continued)

Assertion	Formulation
Composition of CA (Sect. 7.2)	Any complex activity (as well as any of its elements that are a complex activity) can be represented in the form of an aggregate of elements of the following types organized and united by a common goal-setting: • specific elementary operations (representing elementary activity); • control elements of activity (SEAs or elementary operations) that implement analysis, synthesis, specification, regulation and evaluation of CA The links between the elements (their organization) are the nature of the exchange of information messages and / or the exchange of information through a common resource—the CA information model The most "complex" components of management are analysis and synthesis. While the specification (consisting in establishing and maintaining links between the subjects of CA elements, resources and subjects of the subordinate elements) and regulation (constant or multiple, or continuous verification of the occurrence of certain conditions and the initiation of relevant elements of activity) are "simple" and routine
Universal algorithm for control of CA (Sect. 7.2)	The management of any element of complex activity (together with its inferior elements) can be represented by a universal algorithm, which is described by the SEA process model, the goal setting and technology creation model, and the resource pool model (Sects. 5.2, 5.3, 5.4; BPMN diagrams, Figs. 5.3, 5.4 and 5.11)
Aggregated view of CA (Sect. 7.2)	Any complex activity (as well as any of its elements that are a complex activity) can be represented in the form of an aggregate of elements of the following types organized and united by a common goal-setting: • specific elementary operations (representing elementary activity), united by a single logical and single causal structure; • the only control element of the activity that implements analysis, synthesis, specification, regulation and evaluation of CA Obviously, the subject of the aggregated control element will be combined will unite all the subjects of the initial elements of the CA, similarly, the goals, technologies and other components of the CA
Optimization of complex activity (Sect. 7.3)	The task of optimizing complex activity as a whole can be posed and solved in one of two variants: optimizing the performance of CA with known technology and the synthesis of optimal CA technology

(continued)

(continued)

Assertion	Formulation
Optimal realization of CA with known technology (Sect. 7.4)	The tasks of optimal realization of a CA with an approved technology can be formulated as interrelated tasks: • optimal assignment of dissimilar resources taking into account multiple consumer SEA; • optimal maintenance in the required ranges of resource characteristics throughout their CC, primarily during use intervals; • optimal maintenance of the pool of resources in the required ranges.
Synthesis of optimal technology (Sect. 7.5)	The synthesis of technology is essentially specific, and therefore itself allows optimization only to the formation of several alternative options and choosing the best among them Synthesis of technology at the same time generates resource requirements and related optimization tasks: • optimization of the set of resource pools based on the needs of the technologies of various elements of the CA; • optimization of the volumes of each resource pool; • optimizing the maintenance of resource characteristics within specified ranges during their entire LC

Appendix D
Requirements to Methodology of Complex Activity and Its Components

Requirement	Corresponding components of MCA and sections of this book
a. The MCA should include models of both elementary and complex activity, i.e. as having an internal structure, and not possessing. Hence, models of elements of activity and instruments for their integration are needed	Structural element of activity, SEA (Sects. 3.1 and 3.2) Logical structure of MCA (Sect. 3.3) Assertion "Foundations of logical structure of CA" (Sect. 3.3)
b. CA possesses a logical structure, in general a multi-level one. Since CA elements of different levels are themselves CAs, one can say that CA is fractal or self-similar. Structural models of CA must reflect its hierarchy, nesting and fractality	
c. The technology of complex activity determines the cause-effect relationships between the elements of complex activity; hence, MCA should contain cause-effect models of CA	Cause-effect structure of MCA (Sect. 3.4) Assertion "Causal model of CA" (Sect. 3.4)

(continued)

M. V. Belov and D. A. Novikov, *Methodology of Complex Activity*, Studies in Systems, Decision and Control 300, https://doi.org/10.1007/978-3-030-48610-5

(continued)

Requirement	Corresponding components of MCA and sections of this book
d. Complex activity is focused; therefore, MCA should allow to describe and analyze the structure of the goals of CA, as well as the characteristics of the degree of achievement of the goals, the creation of value/utility as a result of CA	Structural element of activity, SEA (Sects. 3.1 and 3.2) Logical structure of MCA (Sect. 3.3) Assertion "Foundations of logical structure of CA" (Sect. 3.3) Definitions and models of Chap. 6
e. Elements of CA exist in time: the needs for the results of CA arise, generate elements of CA, which are realized and then cease to exist. MCA should describe the life cycles of CA elements	Generation models of CA; regular, replicative and creative activity. Chapter 4 Model of forming demand and actualizing need (Sect. 5.1) Process model of SEA (Sect. 5.2) Implementation model of CA as a system (Sect. 5.5) Assertions: "The order of generation of elements of CA" (Sect. 4.1); "External and uncertain nature of demand for CA results" (Sect. 2.8); "Sources of demand for performance results" (Sect. 5.1)
f. MCA should allow describing and analyzing the uncertainty of CA (measurable and true), realized via the occurrence of a priori unpredictable events. The response to uncertainty (the occurrence of events) is the generation of a new activity (absent before the event) with a known technology or a new technology to be created. MCA should describe the generation of new elements of CA	
g. MCA should describe the creation of new technology of activity, as a consequence of requirement f)	Model of setting goals and creating technology for new element of activity (Sect. 5.3)
h. The resources used, consumed and accumulated during the implementation of activity are an essential aspect affecting the technology, subject and subject matter of CA. MCA should contain models for the organization and use of resources	Implementation model of life cycle of resources (Sect. 5.4)
i. MCA should include models of activity such as organization and management	Assertions: "Subject matter of organization and management in relation to CA" (Sect. 1.1 "Essence of organization and management" (Sect. 1.1); "Composition of CA" (Sect. 7.2); "Aggregated view of CA" (Sect. 7.2) Other results and models of Chap. 7
j. Complex activity are implemented in the form of elements, each of which is conventionally related to process or project types, which should be taken into account in the respective models. MCA should combine project and process approaches within a single formalism	Structural element of activity, SEA (Sect. 3.1) Properties of SEAs (Sect. 3.2)

(continued)

(continued)

Requirement	Corresponding components of MCA and sections of this book
k. MCA should represent all modern forms of organization of activity (see Sect. 2.3): (a) elementary forms and (b) complex operations; (c) projects and project programs; (d) life cycles	Structural element of activity, SEA (Sect. 3.1) Properties of SEAs (Sect. 3.2)
l. Multiple and complex links between CA elements and their subjects, the emergence of "meta-subjects" of CA that constitute "meta-organizational systems" (extended enterprises) are essential aspects of CA that should be taken into account within the framework of MCA	Structural element of activity, SEA (Section 3.1) Properties of SEAs (Sect. 3.2) Assertions: "Multiagent representation of CA" (Section 5.5); "Composition of CA" (Sect. 7.2)

References

1. Ackoff, R.: Towards a Systems of Systems Concepts. Manag. Sci. **17**(11), 661–671 (1971)
2. Altshuller, G.: And Suddenly the Inventor Appeared: TRIZ, The Theory of Inventive Problem Solving, p. 171, 2nd edn. Technical Innovation Center, Worchester (1996)
3. Ansoff, H.: Strategic Management, p. 251. Palgrave Macmillan, New York (2007)
4. Aoki, M.: Horizontal vs. Vertical Information Structure of the Firm. Am. Econ. Rev. **76**, 971–983 (1986)
5. Ausenda, G.: On Effectiveness, p. 258. Boydell Press Center for Interdisciplinary Research on Social Stress, Rochester (2003)
6. Barclays Flags « Black Swan Threats » to Commodities This Year. Bloomberg News 5 Jan 2017, 8:49 GMT + 3 https://www.bloomberg.com/news/articles/2017-01-05/barclays-sees-black-swan-threats-to-commodities-in-year-ahead-ixjyd4pq. Accessed 18 Apr 2017
7. Beer, S.: Cybernetics and management, p. 214. The English University Press, London (1959)
8. Belov, M.: General-system modelling framework of complex activity. In: Al-Dabass, D., Colla, V., Vannuci, M., Pantelous, A. (eds.) 10th European Modelling Symposium on Mathematical Modelling and Computer Simulation. IEEE Conference Record No. 40530. Pisa, p. 241 (2016)
9. Belov, M., Kroshilin, A., Repin, V.: ROSATOM's NPP development system architecting: systems engineering to improve plant development. In: Hammami, O., Krob, D., Voirin, J.-L. (eds.) Proceedings of the Second Conference on Complex Design & Management, pp. 255–268. Springer, Berlin (2011)
10. Belov, M., Novikov, D.: Reflexive models of complex activity. In: Proceedings of WOSC World Congress, Rome (2017)
11. Berberian, B., Le Goff K., Rey, A.: Toward a model for effective human-automation interaction. In: Proceedings of 6th International Conference on Digital Human Modeling. Applications in Health, Safety, Ergonomics and Risk Management: Ergonomics and Health, Los Angeles, Part II, pp. 274–283 (2015)
12. Berge, P., Pomeau, Y., Vidal, C.: Order within Chaos, p. 329. Wiley, New York (1987)
13. Bergstra, J.: Process algebra for synchronous communication. Inf. Control **60**, 109–137 (1984)
14. Bertalanffy, L.: General system theory—a critical review. General Syst. **VII**, 1–20 (1962)
15. Bloomberg Tech. Uber Raises Funding at $62.5 Billion Valuation. 3 December 2015. http://www.bloomberg.com/news/articles/2015-12-03/uber-raises-funding-at-62-5-valuation. Accessed 20 Sept 2016
16. Boardman, J., Sauser, B.: Systems Thinking: Coping with 21st Century Problems, p. 240. CRC Press, Boka Raton (2008)
17. Bogdanov, AA.: Algemeine Organisationslehre (Tektologie). Hirzel, Berlin (1926). I; 1928. II / Bogdanov A. Essays in Tektology, p. 291. Intersystems Publications, Seaside (1980)

M. V. Belov and D. A. Novikov, *Methodology of Complex Activity*, Studies in Systems, Decision and Control 300, https://doi.org/10.1007/978-3-030-48610-5

18. Bolton, P., Dewatripont, M.: Contract Theory, p. 740. MIT Press, Cambridge (2005)
19. Broniatowski, D., Moses, J.: Measuring flexibility, descriptive complexity, and rework potential in generic system architectures. Syst. Eng. **19**(3), 207–221 (2016)
20. Bubnicki, ZZ.: Modern Control Theory, p. 423. Springer, Berlin (2005)
21. Burkov,, V., Goubko,, M., Korgin,, N., Novikov,, D.: Introduction to Theory of Control in Organizations, p. 352. CRC Press, New York (2015)
22. Business Process Model and Notation (BPMN), v2.0.2. http://www.omg.org/spec/BPMN/2.0
23. Cannon, WW.: Physiological regulation of normal states: some tentative postulates concerning biological homeostatics. In: Pettit, A. (ed.) A Charles Richet, ses amis, ses collègues, ses élèves (in French), p. 91. LLes Éditions Médicales, Paris (1926)
24. Cannon, WW.: The Wisdom of the Body, p. 294. W.W. Norton & Company, New York (1932)
25. Cecere, LL.: Why Is Sales and Operations Planning So Hard? FORBES. Logistics & Transportation. 1/21/2015 @ 2:35PM 8 285 views http://www.forbes.com/sites/loracecere/2015/01/21/why-is-sales-and-operations-plannning-so-hard. Accessed 17 Apr 2017
26. Charnes, A., Cooper, W., Rhodes, E.: Measuring the efficiency of decision making units. Eur. J. Oper. Res. **2**, 429–444
27. Checkland,, P.: Systems Thinking, Systems Practice, p. 424. Wiley, New York (1999)
28. CIMdata PLM Glossary. http://www.cimdata.com/en/resources/about-plm/cimdata-plm-glossary#PLM. Accessed 30 Jan 2017
29. CIMdata Publishes PLM Market and Solution Provider Report. Wednesday, July 06, 2016. http://www.cimdata.com/en/news/item/6459-cimdata-publishes-plm-market-and-solution-provider-report. Accessed 30 Jan 2017
30. Dalkir, KK.: Knowledge Management in Theory and Practice, p. 485. 2nd edn. MIT Press, Cambridge (2011)
31. Davenport, TT., Prusak,, L.: Working Knowledge: How Organizations Manage what they Know, p. 199. Harvard Business Scholl Press, Cambridge (1998)
32. De Smet, AA., Lund, SS., Schaninger,, W.: Organizing for the Future. Platform-based Talent Markets Help Put the Emphasis in Human-capital Management Back where it Belongs-on Humans, January 2016 http://www.mckinsey.com/insights/organization/organizingffort theffuture. Accessd 15 Sept 2016
33. DeRosa, JJ., Grisogono, AA., Ryan, AA., Norman,, D.: A Research agenda for the engineering of complex systems. In: Proceedings of IEEE International Systems Conference, Montreal, pp. 1–8 (2008)
34. Dewar, JJ.: Assumption Based Planning a Tool for Reducing Avoidable Surprises, p. 248. Cambridge University Press, Cambridge (2002)
35. Dorf, R., Bishop, R.: Modern Control Systems, p. 1111. 12th edn. Prentice Hall, Upper Saddle River (2011)
36. Dori, D., Sharon, A.: Integrating the project with the product for applied systems engineering management. In: 14th IFAC Symposium on Information Control Problems in Manufacturing. IFAC Proceedings Volumes, vol. 45. Issue 6, pp. 1153–1158 (2012)
37. Duhon, B.: It's all in our heads. Information **12**(8), 8–13 (1998)
38. Namatame, A., Kurihara, S., Nakashima, H.: Emergent Intelligence of Networked Agents, p. 261. Springer, Berlin (2007)
39. Engeström, Y.: Learning by Expanding, p. 426. Orienta Konsultit, Helsinki (1987)
40. Engeström, Y.: Learning by Expanding: An Activity-Theoretical Approach to Developmental Research, p. 338, 2nd edn. Cambridge University Press, Cambridge (2014)
41. Engeström, Y.: The future of activity theory: a rough draft. In: Sannino, A., Daniels, H., Gutierrez, K. (eds.) Learning and Expanding with Activity Theory, p. 367. Cambridge University Press, Cambridge (2009)
42. Estefan, J.: Survey of Candidate Model-Based Systems Engineering (MBSE) methodologies. In: International Council on Systems Engineering (INCOSE). INCOSE-TD-2007-003-02. http://www.omgsysml.org/MBSE_Methodology_Survey_RevB.pdf. Accessed 17 Dec 2016
43. Farrell, M.: The Measurement of Productive Efficiency. J. R. Statist. Soc. Ser A (General). Part III **120**, 253–281 (1957)

44. Foss, N., Lando, H., Thomsen, S.: The Theory of the Firm. In: Bouckaert, B., De Geest, G. (eds). Encyclopedia of Law and Economics. Volume III. The Regulation of Contracts, pp. 631–658. Edward Elgar, Cheltenham (2000)
45. Galloway, L.: Principles of Operations Management, p. 235. Cengage Learning EMEA, London (1998)
46. Germeier, Y.: Non-antagonistic games, 1976, p. 331. D. Reidel Publishing Company, Dordrecht (1986)
47. Gilbert, D., Yearworth, M.: Complexity in a systems engineering organization: an empirical case study. Syst. Eng. **19**(5), 422–435 (2016)
48. Gorod, A., Gandhi, S., White, B., Ireland, V., Sauser, B.: Modern history of system of systems, enterprises, and complex systems. In: Case Studies in System of Systems, Enterprise Systems, and Complex Systems Engineering, pp. 3–32. CRC Press, Boca Raton (2014)
49. Gorod, A., Gandhi, S., White, B., Ireland, V., Sauser, B.: Relevant aspects of complex systems from complexity theory. Case Studies in System of Systems, Enterprise Systems, and Complex Systems Engineering, pp. 33–78. CRC Press, Boca Raton (2014)
50. Gralla, E., Szajnfarber, Z.: Characterizing representational uncertainty in system design and operations. Syst. Eng. **19**(6), 535–548 (2016)
51. Grisogono, A.: The implications of complex adaptive systems theory for C2. In: State of the Art and the State of the Practice, CCRTS 2006, p. 19. Department of Defence Command & Control Research Program, San Diego (2006)
52. Grossman, S., Hart, O.: The costs and benefits of ownership: a theory of vertical and lateral integration. J. Polit. Econ. **94**(4), 691–719 (1986)
53. Hamme, M., Champy, J.: Reengineering the Corporation: A Manifesto for Business Revolution, p. 223. Harper Business, New York (1993)
54. Hammond, J.: Learning by the Case Method. Harvard Business School. Rev. 16 April 2002. http://isites.harvard.edu/fs/docs/icb.topic1236622.files/CaseMethod.pdf
55. Hoare, C.: Communicating sequential processes. CACM **21**(8), 666–677 (1978)
56. Hoare, C.: Communicating Sequential Processes, p. 256. Prentice Hall, New York (1985)
57. Hodgson, G., Knudsen, T.: Generative replication and the evolution of complexity. J. Econ. Behav. Organ. **75**, 12–24 (2010)
58. Hodgson, G., Knudsen, T.: Why we need a generalized darwinism and why generalized darwinism is not enough. J. Econ. Behav. Organ. **61**, 1–9 (2006)
59. Holland, J.: J. Syst. Sci. Complex. **19**(1), 1–8 (2006)
60. http://www.valuebasedmanagement.net. Accessed 24 Dec 2016
61. https://en.wikipedia.org/wiki/Resource. Accessed 24 Dec 2016
62. Haskins C. (ed.).: INCOSE Systems Engineering Handbook Version 3.2.2—A Guide for Life Cycle Processes and Activities, p. 376. INCOSE, San Diego (2012)
63. ISO 55000:2014(en) Asset Management—Overview, Principles and Terminology
64. ISO 9000:2000 Quality Management Systems—Fundamentals and Vocabulary
65. ISO/IEC 16085:2006 Systems and Software Engineering—Life Cycle Processes—Risk management
66. ISO/IEC FDIS 16085 IEEE Std 1540-2006 Systems and Software Engineering—Life Cycle Processes—Risk Management
67. ISO/IEC/IEEE 15288:2015 Systems and Software Engineering—System Life Cycle Processes
68. ISO/IEC/IEEE 42010:2011 Systems and Software Engineering—Architecture Description
69. Jackson, M.: Systems Thinking—Creative Holism for Managers, p. 378. Wiley, Chichester (2003)
70. Knight, F.: Risk, Uncertainty and Profit/Hart Schaffner and Marx Prize Essays. No. 31, p. 381. Houghton Mifflin, Boston and New York (1921)
71. Kuhn, T.: The Structure of Scientific Revolutions, p. 264. University of Chicago Press, Chicago (1962)
72. Kurzweil, R.: How to Create a Mind: The Secret of Human Thought Revealed, p. 352. Viking Books, New York (2012)

73. Kurzweil, R.: The Singularity is Near: When Humans Transcend Biology, p. 672. Viking Books, New York (2005)

74. Lakatos, I.: The Methodology of Scientific Research Programmes: Volume 1: Philosophical Papers, p. 250. Cambridge University Press, Cambridge (1978)

75. Lefevbre, V.: The Structure of Awareness: Toward a Symbolic Language of Human Reflexion, p. 199. Sage Publications, New York (1977)

76. Leontiev, A.: Activity, Consciousness and Personality, p. 148. Prentice-Hall, Upper Saddle River (1979)

77. Mankins, J.: Technology Readiness Levels: A White Paper / 6 April 1995 NASA Office of Space Access and Technology, Advanced Concepts Office. http://www.hq.nasa.gov/office/codeq/trl/trl.pdf. Accessed 27 Dec 2016

78. Mann, S.: Chaos theory in strategic thought. Parametes **XXII**, 54–68 (1992)

79. ManuCloud: The Next-generation Manufacturing as a Service (MaaS) Environment. http://www.manucloud-project.eu/index.php?id=233. Accessed 20 May 2015

80. Marshal, A.: Principles of Economics. Macmillan and Co., London. http://oll.libertyfund.org/titles/marshall-principles-of-economics-8th-ed. Accessed 10 May 2017

81. Mason, R., Mitroff, I.: Challenging Strategic Planning Assumptions, p. 324. Wiley Inc, New York (1981)

82. Massaki, I.: Kaizen. Key to Japan's Competitive Success. Random House, New York (1988). 256 p

83. Maturana, H., Varela, F.: The Cognitive Process. Autopoiesis and Cognition: The Realization of the Living, p. 171. D. Reidel Publishing Company, Dordrecht (1980)

84. Mauldin, J.: 10 Potential Black Swans and Opportunities for the US Economy in 2017 // FORBES. Jan 6, 2017 @ 06:31 PM. https://www.forbes.com/sites/johnmauldin/2017/01/06/10-potential-black-swans-and-opportunities-for-the-us-economy-in-2017/#53af8f5a174e. Accessed 18 Apr 2017

85. McNair, M., Hersum, A.: The Case Method at the Harvard Business School: Papers by Present and Past Members of the Faculty and Staff, p. 292. McGraw-Hill, New York (1954)

86. Mechanism Design and Management: Mathematical Methods for Smart Organizations/Ed. by Prof. Novikov D., p. 163. Nova Science Publishers, New York (2013)

87. Mesarović, M., Mako, D., Takahara, Y.: Theory of Hierarchical Multilevel Systems, p. 294. Academic, New York (1970)

88. Mescon, M., Albert, M., Khedouri, F.: Management, p. 777, 3rd edn. Harpercollins College Div, New York (1988)

89. Milgrom, P., Roberts, J.: The economics of modern manufacturing: technology, strategy, and organization. Am. Econ. Rev. **80**, 511–528 (1990)

90. Milner, R.A.: Calculus of Communicating Systems. Lecture Notice in Computer Science, p. 171, vol. 92. Springer Verlag, Heidelberg (1980)

91. Mintzberg, H.: Structure in Fives: Designing Effective Organizations, p. 312. Prentice-Hall, Englewood Cliffs, NJ (1983)

92. Mishin, S.: Optimal Hierarchies in Firms, p. 127. PMSOFT, Moscow (2004)

93. Mitter, S.: On systems effectiveness. In: Ausenda, G. (ed.) On Effectiveness. Center for Interdisciplinary Research on Social Stress, pp. 37–48. Boydell Press, Rochester (2003)

94. Mosleh, M., Ludlow, P., Heydari, B.: Distributed resource management in systems of systems: an architecture perspective. Syst. Eng. **19**(4), 362–374 (2016)

95. Nell, J.: An Overview of GERAM. In: 1997 International Conference on Enterprise Integration Modelling Technology, ICEIMT'97. Updated 30 January 1997. https://web.archive.org/web/19990221190103/, http://www.mel.nist.gov/workshop/iceimt97/ice-gera.htm (1997). Accessed 23 June 2017

96. North, M.A.: Theoretical formalism for analyzing agent-based models. Complex Adap. Syst. Model. (2014). https://doi.org/10.1186/2194-3206-2-3, http://link.springer.com/article/10.1186/2194-3206-2-3. Accessed 21 Dec 2016

97. Novikov, A., Novikov, D.: Methodology, p. 668. Sinteg, Moscow (2007) (in Russian)

98. Novikov, A., Novikov, D.: Research Methodology: From Philosophy of Science to Research Design, p. 130. CRC Press, Amsterdam (2013)
99. Novikov, D.: Control Methodology, p. 76. Nova Science Publishers, New York (2013)
100. Novikov, D.: Cybernetics: From Past to Future, p. 107. Springer, Heidelberg (2016)
101. Novikov, D.: Theory of Control in Organizations, p. 341. Nova Science Publishers, New York (2013)
102. Oracle. http://www.oracle.com/ru/index.html. Accessed 26 Feb 2017
103. Page, S.: Understanding Complexity. The Great Courses. Chantilly: The Teaching Company, p. 224 (2009)
104. Palmatier G., Crum C. Enterprise Sales and Operations Planning: Synchronizing Demand, Supply and Resources for Peak Performance. – N.Y.: J. Ross Publishing, 2002. – 288 p
105. Palmatier, G., Crum, C.: The Transition from Sales and Operations Planning to Integrated Business Planning, p. 112. Oliver Wight Int, New London (2013)
106. Peregudov, F., Tarasenko, F.: Introduction to Systems Analysis, p. 320. Glencoe/Mcgraw-Hill, OH: Columbus (1993)
107. Pfanzagl, J.: Theory of Measurement, p. 240, 2nd edn. Physica, London (1971)
108. Popper, K.: The Logic of Scientific Discovery, p. 513. Routledge, London (1959)
109. Porter, M.: Competitive Advantage: Creating and Sustaining Superior Performance, p. 560. Collier Macmillan, London (1985)
110. Porter, M.: Towards a Dynamic Theory of Strategy. Strateg. Manag. J. **12**, 95–117. Special Issue (1991)
111. Porter, M., Heppelmann, J.: How Smart, Connected Products Are Transforming Companies, Harvard Business Review, p. 19, October 2015, REPRINT R1510G
112. Poston, T., Stewart, I.: Catastrophe Theory and Its Applications, p. 491. Dover Publications, New York (1997)
113. Prigogine, I., Stengers, I.: Order Out of Chaos, p. 285. Bantam Books, New York (1984)
114. In: Proceedings of the First International Workshop on Activity Theory Based Practical Methods for IT-Design ATIT-2004. 2-3. September 2004, Copenhagen, Denmark
115. Rashevsky, N.: Outline of a New Mathematical Approach to General Biology. Bull. Math. Biophys. **5**, 33–47, 49–64, 69–73 (1943)
116. Rebovich, G., White, B.: Enterprise Systems Engineering: Advances in the Theory and Practice, p. 459. CRC Press, Boca Raton (2011)
117. Ries, E.: The Lean Startup: How Constant Innovation Creates Radically Successful Businesses, p. 336. Portfolio Penguin, New York (2011)
118. Rosser, J.: Complexities of natural selection dynamics. In: Global Analysis of Dynamic Models in Economics and Finance, pp. 429–442. Springer, Heidelberg (2013)
119. Rubinstein, S.: Probleme Der Allgemeinen Psychologie, p. 488. Springer-Verlag, New York (1981)
120. Rzevski, G., Skobelev, P.: Managing Complexity, p. 216. WIT Press, London (2014)
121. Salanie, B.: The Economics of Contracts, p. 224, 2nd edn. MIT Press, Massachusetts (2005)
122. http://www.sap.com/cis/index.html. Accessed 26 Feb 2017
123. Satzinger, J., Jackson, R., Burd, S.: Introduction to Systems Analysis and Design, p. 512, 6th edn. Course Technology, Boston (2011)
124. Savin, S.: The use of modern polymer composite materials in the glider design of the MS-21 Aircraft. In: Proceedings of the Samara Scientific Center of the Russian Academy of Sciences , vol. 14, n 4(2). pp. 686 – 693 (2012) (in Russian)
125. Schumpeter, J.: The Theory of Economic Development, p. 255. Harvard University Press, Cambridge (1934)
126. Schwab, K.: The Fourth Industrial Revolution, p. 192. Portfolio Penguin, New York (2017)
127. Sheard, S., Mostashari, A.: Principles of complex systems for systems engineering. Syst. Eng. **12**(4), 295–311 (2009)
128. Shoham, Y., Leyton-Brown, K.: Multiagent Systems: Algorithmic, Game-Theoretic, and Logical Foundations. Cambridge University Press, Cambridge, p. 504 (2008)
129. Simon, H.: The Architecture of Complexity. Proc. Am.Philos. Soc. **106**(6), 467–482 (1962)

130. Simon, H.: The Sciences of the Artificial, p. 241, 3rd edn. MIT Press, Cambridge (1969)
131. Sinha, K., Weck, O.: Empirical validation of structural complexity metric and complexity management for engineering systems. Syst. Eng. **19**(3), 193–206 (2016)
132. Stole, L.: Lectures on the Theory of Contracts and Organizations, p. 104. University of Chicago, Chicago (1997)
133. Stone, P., et al.: Artificial Intelligence and Life in 2030. One Hundred Year Study on Artificial Intelligence: Report of the 2015-2016 Study Panel. Stanford University, Stanford. http://ai100. stanford.edu/2016-report (2016). Accessed 01 Aug 2016
134. Sutherland, J.: Scrum: The Art of Doing Twice the Work in Half the Time, p. 256. Random House Business, New York (2015)
135. Systems and Software Engineering Vocabulary (SEVocab) – ISO/IEC 24765. In: International Organization for Standardization (ISO) / International Electrotechnical Commission (IEC) [database online]. Geneva, Switzerland. http://pascal.computer.org/sev_display/index. action(2009). Accessed 03 Mar 2017
136. Systems Architecture. MIT Course Number ESD.34 MITOPENCOURSWEAR, Massachusetts Institute of Technology. https://ocw.mit.edu/courses/engineering-systems-div ision/esd-34-system-architecture-january-iap-2007. Accessed 03 Mar 2017
137. Systems Engineering Guide, p. 710. MITRE Corporation, Bedford (2014)
138. Taha, H.: Operations Research: An Introduction, p. 813, 9th edn. Prentice Hall, New York (2011)
139. Taleb, N.: Antifragile: Things that Gain from Disorder, p. 544. Random House, New York (2012)
140. Taleb, N.: The Black Swan: The Impact of the Highly Improbable, p. 444, 2nd edn. Random House Trade Paperbacks, New York (2011)
141. Technology Readiness Assessment (TRA) Guidance. United States Department of Defense, April 2011. http://www.acq.osd.mil/chieftechnologist/publications/docs/TRA2011. pdf. Accessed 27 Dec 2016
142. Levine, W.: The Control Handbook, p. 786, 2nd edn. CRC Press, New York (2010)
143. Kroszner, R., Putterman, L (eds.).: The Economic Nature of the Firm. A Reader, p. 390, 2nd edn. Cambridge University Press, Harvard (1996)
144. The Guide to the Systems Engineering Body of Knowledge (SEBoK) v. 1.2 / Pyster, A., Olwell, A. (eds.). The Trustees of the Stevens Institute of Technology, Hoboken. http://www. sebokwiki.org (2013). Accessed 20 Dec 2016
145. The Open Group: The Open Group Architecture Framework (TOGAF) – Version 9, Enterprise Edition. http://pubs.opengroup.org/architecture/togaf9-doc/arch (2009). Vizited 23 июня 2017
146. The World's Biggest Public Companies. www.forbes.com/global2000. Accessed 26 Feb 2017
147. UK Department of Transport TRAK Steering Group: TRAK Enterprise Architecture Framework. http://trak.sourceforge.net (2010) . Accessed 23 June 2017
148. US Department of Defense: DoD Architecture Framework Version 2.0. http://dodcio.defense. gov/Portals/0/Documents/DODAF/DoDAF_v2-02_web.pdf. (2009). Accessed 23 June 2017
149. US Department of the Treasury Chief Information Officer Council (2000). Treasury Enterprise Architecture Framework. Version 1, July 2000. https://ru.scribd.com/document/63363136/Tre asury-Enterprise-Architectu-re-Framework-TEAF. Accessed 23 June 2017
150. Wagner, H.: Principles of Operations Research, p. 1039, 2-nd edn. Prentice Hall, NJ Upper Saddle River (1975)
151. Warfield, J.: An Introduction to Systems Science, p. 403. World Scientific Publishing, London (2006)
152. Weaver, W.: Science and Complexity. Am. Sci. **36**, 536–544 (1948)
153. Weinberg, G.: An Introduction to General Systems Thinking, p. 304. Dorset House, New York (2001)
154. White, B.: Fostering Intra-Organizational Communication of Enterprise Systems Engineering Practices. MITRE Public Release Case No. 06-0351. National Defense Industrial Association, 9th Annual Systems Engineering Conference. San Diego, p. 25 (2006)

155. Wooldridge, M.: An Introduction to Multi-Agent Systems, p. 376. Wiley, New York (2002)
156. Yanhui, W.: Organizational structure and product choice in knowledge-intensive firms. Manag. Sci. **61**(8), 1830–1848 (2015)
157. Zachman, J.: A framework for information systems architecture. IBM Syst. J. **26**, 276–292 (1987)

Printed in the United States
by Baker & Taylor Publisher Services